LF

Protected Agriculture

A Global Review

Merle H. Jensen and Alan J. Malter

The World Bank
Washington, D.C.

Technical Papers are published to communicate the results of the Bank's work to the development community with the least possible delay. The typescript of this paper therefore has not been prepared in accordance with the procedures appropriate to formal printed texts, and the World Bank accepts no responsibility for errors. Some sources cited in this paper may be informal documents that are not readily available.

The findings, interpretations, and conclusions expressed in this paper are entirely those of the author(s) and should not be attributed in any manner to the World Bank, to its affiliated organizations, or to members of its Board of Executive Directors or the countries they represent. The World Bank does not guarantee the accuracy of the data included in this publication and accepts no responsibility whatsoever for any consequence of their use. The boundaries, colors, denominations, and other information shown on any map in this volume do not imply on the part of the World Bank Group any judgment on the legal status of any territory or the endorsement or acceptance of such boundaries.

The material in this publication is copyrighted. Requests for permission to reproduce portions of it should be sent to the Office of the Publisher at the address shown in the copyright notice above. The World Bank encourages dissemination of its work and will normally give permission promptly and, when the reproduction is for noncommercial purposes, without asking a fee. Permission to copy portions for classroom use is granted through the Copyright Clearance Center, Inc., Suite 910, 222 Rosewood Drive, Danvers, Massachusetts 01923, U.S.A.

The complete backlist of publications from the World Bank is shown in the annual *Index of Publications*, which contains an alphabetical title list (with full ordering information) and indexes of subjects, authors, and countries and regions. The latest edition is available free of charge from the Distribution Unit, Office of the Publisher, The World Bank, 1818 H Street, N.W., Washington, D.C. 20433, U.S.A., or from Publications, The World Bank, 66, avenue d'Iéna, 75116 Paris, France.

ISSN: 0253-7494

Merle H. Jensen is professor of Plant Sciences at the University of Arizona. At the time this paper was prepared, Alan J. Malter was a market researcher for the Ministry of Agriculture in Israel.

Library of Congress Cataloging-in-Publication Data

Jensen, Merle H.
    Protected agriculture : a global review / Merle H. Jensen and
Alan J. Malter.
        p.   cm. — (World Bank technical paper, ISSN 0253-7494 ; no.
253)
    Includes bibliographical references (p.   ).
    ISBN 0-8213-2930-8
    1. Greenhouse management. 2. Plants, Protection of.
3. Floriculture. 4. Truck farming.   I. Malter, Alan J.   II. Title.
III. Series.
SB415.J46   1994
635'.0483—dc20   94-24839
                                                      CIP

TABLE OF CONTENTS

PART 3   PRODUCTION ASPECTS

FOREWORD

As land and water resources become increasingly constrained for agriculture in many parts of the world and in the urban areas in particular, there has been a rapid upsurge in the production of high value crops under plastic and glass. The removal of trade barriers, coupled with growing consumer demand for quality produce all year round, has further stimulated this move towards high value, intensive forms of horticultural production. Importantly, protected cultivation, while requiring farmers to have a more comprehensive knowledge of agronomic and crop management principles than for traditional agriculture, is well adapted to both small and large scale commercial operations.

The success of protected agriculture is very much dependent on the level and quality of applied technology and on factors such as the local climate, the buying power of the consumer, transport organization, market intelligence and access to local and international markets. An important aspect for succesful exports is the time of the year that fresh vegetables and flowers can be made available to fit in a niche in the overseas markets at a time when local production is not sufficient or absent altogether. Producers in arid and cool climates in the Southern hemisphere and the Middle East have therefore an advantage in dealing with Northern markets.

In response to the need to be better informed of the technical developments in protected cultivation and their application through World Bank supported projects, the Agriculture and Natural Resources Department has prepared this technical publication, primarily for use by World Bank staff and those responsible for implementing such projects. The paper draws together globally applied technologies, best practices and international markets experience as reflected in the special annex to this paper.

Alexander F. McCalla

Director Agriculture and
Natural Resources Department

ABSTRACT

This Paper, "Protected Agriculture, A Global Review", gives a broad overview of the three main protective methods of plant coverage and related technologies for climate control and production techniques, including drip irrigation in vegetables and floriculture. The three protective methods are by using mulches, row covers and green houses. The Paper also addresses some relevant economic and policy aspects in protected agriculture. Because of the overriding importance of the sector, a special ANNEX on modern marketing has been included to highlight world wide marketing issues and trends from centers in Europe, North America, the Middle East, to Latin America. The Paper is generously illustrated and written specifically for World Bank's operational staff and national counterparts in developing countries.

The Paper is organized in four PARTS and a special ANNEX on Marketing. PART I gives the Background in Chapters 1 and 2, PART II deals with the Protective Materials and Structures in Chapters 3 and 4, PART III discusses the main Production Aspects in Chapters 5 to 10, while PART IV highlights the Economic Considerations in the remaining Chapters 11 to 14. The ANNEX on Marketing has been written by a different author and can be read seperately from Chapter 12 on Marketing and Distribution in PART IV of the document. However because of the complementarity of the contents, it is highly recommended to read both the ANNEX and Chapter 12.

ACKNOWLEDGEMENTS

The publication of this Technical Paper, "Protected Agriculture, A Global Review" was made possible under an AGR Study, which was initially managed by Hamdy Eisa, Principle Agriculturist. The need to address the subject is based on the Bank's increased attention to agricultural intensification and diversification issues, and particularly to high value crop production under glass and plastic by smallholders in arid and cool climates. The main author is Merle H. Jensen, while a special Annex on Marketing was written by Alan J. Malter.

Dr. Merle H. Jensen is Professor of Plant Sciences, Associate Director of the Agricultural Experiment Station and Assistant Dean for Sponsored Research in the College of Agriculture at the University of Arizona. He is a world authority on protected agriculture and has written many research and technical papers on the subject.

Mr. Alan J. Malter is a market researcher for horticultural products in the Market Research Department of the Ministry of Agriculture, Israel. His contribution to the Paper deals with the marketing and distribution of protected crops. It examines the broad trends in the USA, European and Japanese consumption and trade in fresh vegetables and cut flowers, noting the opportunities for "off-season", developing country suppliers.

The first draft reports produced were reviewed and edited by external reviewers and their comments incorporated. Final editing was done by World Bank staff Steven Jaffee, Agro-business Specialist and Johannes ter Vrugt, Senior Agronomist.

The layout and design was carried out by Peter Wiant, graphic consultant.

PART ONE

BACKGROUND

CHAPTERS 1 - 2

1
INTRODUCTION

Protected agriculture is the modification of the natural environment to achieve optimum plant growth. Modifications can be made to both the aerial and root environments to increase crop yields, extend the growing season and permit plant growth during periods of the year not commonly used to grow open field crops. Protected agriculture may also indicate comprehensive systems of controlled environmental agriculture (CEA) in which all aspects of the natural environment are modified for maximum plant growth and economic return. Control may be imposed on air and root temperatures, light, water, humidity, carbon dioxide and plant nutrition, along with complete climatic protection and may ultimately make it possible to grow crops on a lunar base or another planet.

Protected agriculture has enabled many countries to greatly extend their food production capability. Until now, such systems of agriculture have been largely concentrated in developed countries, but recent research developments have made it possible to extend the benefits of this technology to less affluent regions of the world. Plastic greenhouses, row covers, and mulching systems were first used widely in Southern Europe, Japan, and the United States; they are now found in other areas such as the People's Republic of China, South Korea, Middle East and North Africa.

Protected agriculture will play an important role in meeting the world's food production requirements for the year 2025. It is estimated that, by that year, 8.5 billion people will have to be fed - an addition of more than 3 billion people to the world population in 1990. In the developing world alone, the population is expected to reach 5 billion by the year 2025. Already, about half of the Third World is 16 years of age or younger. By the year 2000, it is estimated that half the world's population will live in cities. Of the 24 cities expected to exceed 10 million inhabitants, 18 will be in the less developed countries.

It is clear that if we are to increase the supply of food during the next century, we must increase the output of the land. New farming techniques, which incorporate developments in irrigation, fertilizers, pesticides, herbicides and genetics have already increased production. More disease-resistant crop cultivars have been bred which are responsive to such farm inputs as fertilizer and plant density. The well-known Green Revolution is an example of how modern technology has substantially increased yields in some developing countries with severe food/population problems. For much of the world, climatic conditions have prohibited year-around farming with crop production often limited to only one season. Protected agriculture can provide new alternatives and economic opportunities in crop production to feed a hungry world.

This publication describes the potential and application of protected agriculture for use in developing countries. It details technological advances used throughout the world to increase crop yields, to improve quality and to permit early and out-of-season cropping of food and ornamental crops.

DEFINITION OF TERMS

Many methods of protected agriculture are used to modify the environment. Some are quite simple and inexpensive; others are complex and costly.

Ideally, crop production would take place in an area not requiring protected agriculture systems, an area with year-around growing temperatures. Because such regions are rare, and often completely lacking in most countries, methods have been devised to protect crops against the harsh climatic extremes.

Mulching. Mulching is the practice of covering the soil around plants with an organic or synthetic material to make conditions more favorable for plant growth, development and crop production (Hopen and Oebker 1976).

Row Covers. A row cover is a piece of clear plastic stretched over low hoops and secured along the sides of the plant row by burying the edges and ends with soil. Floating row covers are wide sheets of clear, perforated, polyethylene or non-woven porous plastic, not supported by hoops but by the plant itself.

Two hectare greenhouse facility in United Arab Emirates allowed production of horticultural crops under severe climatic conditions.

High Tunnel. A greenhouse-like unit but without mechanical ventilation or a permanent heating system.

Intensive Agriculture. This is based on cropping systems that may require large amounts of labor and capital per unit area of land per year and normally involves crops of high value (Dalrymple 1973). Such agricultural endeavors usually involve the extension of the growing season by means of protected agriculture. Multiple cropping, the growing of more than one crop per unit of land per year, is common.

Greenhouses. A greenhouse is a framed or inflated structure, covered by a transparent or translucent material that permits optimum light transmission for plant production and protects against adverse climatic conditions. Such a structure enables a person to work inside; and may include mechanical equipment for heating and cooling.

Cucumbers grown in a sand culture system with drip irrigation in Arizona. Cucumber yields in this system exceeded 700 MT/ha.

The three major methods of protected agriculture are: (1) mulches, (2) row covers, and (3) greenhouses. The many systems of choice included in each method are discussed in PART II, Chapters III and IV.

Controlled Environment Agriculture (CEA). While similar to greenhouse agriculture, CEA systems of protected agriculture are the ultimate in environment control. CEA means control at both the aerial and root levels. Production may take place in a greenhouse or in a totally enclosed structure that permits control of air and root temperature, humidity, atmospheric gas composition, light, water, growing medium, and plant nutrition.

<u>**Hydroponics.**</u> Hydroponics is a technology for growing plants in a nutrient solution (water and fertilizers) with or without the use of an artificial medium (e.g. sand, gravel, vermiculite, rock wool, peat moss, sawdust) to provide mechanical support. *Liquid* hydroponic systems have no other supporting medium for the plant roots; *aggregate* systems have a solid medium of support. Hydroponics systems are further categorized as *open* (i.e.,once the nutrient solution is delivered to the plant roots, it is not reused) or *closed* (i.e., surplus solution is recovered, replenished, and recycled).

Some regional growers, agencies, and publications persist in limiting the definition of hydroponics to liquid systems only. This exclusion of aggregate hydroponics blurs the statistical data and may result in an underestimation of the full extent of the technology. This underestimation then distorts the economic implications of hydroponics.

Virtually all hydroponics systems are enclosed in greenhouse-type structures in order to provide temperature control, to reduce evaporative water loss, to better control disease and pest infestations, and to protect hydroponics crops against the elements, such as wind and rain. Thus, while hydroponics and CEA are not synonymous, CEA usually accompanies hydroponics.

Drip Irrigation. Often referred to as trickle irrigation, drip irrigation systems consist of plastic tubes of small diameter laid on the surface or in the subsurface of the field alongside or beneath the plants. Water is delivered to the plants at frequent intervals from small holes or emitters located along the tube, normally at rates between one and six liters of water per hour.

Technical Production Aspects are being discussed in PART III, Chapters V to X.

ECONOMIC RELATIONSHIPS

Crop production, using systems of protected agriculture, is usually more expensive per unit of product than production without such systems during comparable periods of the year. These additional costs are usually justified if the monetary return perunit of product is higher; this occurs when the product is of better quality, if overall production costs are compensated through better yields or if crop production occurs when local cropping is impossible. Calculating the economic advantage is more complicated when imports from other producing areas are considered. For example, the cost of heating fuel required to produce a kilo of greenhouse tomatoes in New York State during the winter is far greater than the cost of fuel to transport to New York a kilo of tomatoes grown in the open field in Mexico at that time.

Economic factors are key determinants of the method and system of protected agriculture most applicable for crop production in a given situation. If early harvest of one to two weeks is desired to take advantage of high market prices,then mulches alone may provide that margin of time and profit. If an advantage of several weeks is desired, a combination of mulches and row covers may be required. Naturally, the cost of using both methods in combination are higher than using mulches alone; however higher market prices might more than compensate for the additional cost. In Japan, all three methods are commonly used simultaneously to grow crops during winter and early spring. In most cases, drip irrigation is also used with all other methods of protected agriculture.

While nearly all crops can be grown successfully in any system of protected agriculture, only those crops which bring a high yield per unit area of land, as well as high market prices, are economically viable. Usually, these are perishable crops that cannot be transported long distances without expensive packaging and shipping costs.

In addition, each method demands specialized knowledge and presents its own economic risks. For instance, greenhouse crop production, in combination with hydroponic culture, is possibly the most intensive method of crop production in today's agricultural industry. It requires high technology and is very capital intensive. Every detail of crop production requires close attention. But excellent management skills in protected agriculture are not enough: a thorough knowledge of markets and the ability to produce high quality products on schedule are also essential.

Economic problems come from many causes: underestimating the cost of production, including capital and operating cost requirements; poor management and marketing skills; and lack of diversification in response to competition from less expensive imported products. Economic success is highly dependent on making protected agricultural methods of production competitive with systems of open field agriculture not using any method of crop protection, whether produced locally or imported.

Economic and marketing factors are broadly discussed in PART IV, from Chapter XI onwards.

GEOGRAPHIC CONSIDERATIONS

When considering the use of any system of protected agriculture, the world may be divided into three geographic regions:

(1) temperate, (2) semi-arid/arid, and (3) tropical. In the temperate regions, all methods of protected agriculture are often used for early crop production and to produce summer crops out-of-season, during the winter. In the temperate regions, mulches add warmth to the root area; in tropical regions, mulches protect fruits from the disease or discoloration that might occur from contact with the soil.

Row covers are commonly used during early spring in both the temperate and arid regions, but are seldomly used in the tropics. One exception to use in the tropics might be the introduction of non-woven materials as protection against chewing insects or insects which are vectors of plant viruses.

Greenhouse structures are enclosed to provide temperature control and opened only to provide ventilation in both temperate and arid regions. In arid regions, during summer and even winter, evaporative cooling systems are commonly used to lower greenhouse temperatures. Closure also provides valuable protection from disease and pest infestations, and weather damage. Greenhouses are especially effective in tropical regions, for these reasons. In the tropics, the sides of a greenhouse structure are often left open for natural ventilation but, if pest infestation is threatening, the sides are covered with screens.

Soilless/hydroponics culture is commonly used in combination with greenhouses, especially where no suitable soil exists and for more efficient use of water and fertilizer. Hydroponics is also used for optimum control over disease and insects.

The major attributes of hydroponics systems will be discussed in Chapter IV.

2
HISTORY

The earliest protected agriculture was possibly the growing of off-season cucumbers under "transparent stone" for the Roman Emperor Tiberius during the first century. The technology was rarely employed, if at all, during the following 1500 years.

During the 1600's, several techniques were used to protect horticultural crops against the cold. These included glass lanterns, bell jars, cold frames and hot beds covered with glass. In the seventeenth century, low portable wooden frames covered with an oiled translucent paper were used to warm the plant environment much as plastic row covers do today (Dalrymple 1973). In Japan, straw mats were used in combination with oil paper to protect crops from the severe natural environment (Takakura 1988). Greenhouses in France and England during the same century were heated by manure and covered with glass panes (Gibault 1912). The first glasshouse in the 1700's, used glass on one side only, as a sloping roof. Later in the century, glass was used on both sides. This glass-house was used for fruit crops such as melons, grapes, peaches, and strawberries and only rarely for vegetable production (Dalrymple 1973). It would seem that the developers of this new technology kept market profitability in mind: they produced crops which appealed to the wealthy and privileged, the only people who could afford the luxury of fresh fruit produced out of season in greenhouses.

Clear polyethylene mulch, in combination with drip irrigation, is used extensively in Southern California, to produce high quality strawberries.

Protected agriculture was fully established with the introduction of polyethlene after World War II. The first use of polyethylene as a greenhouse cover was in 1948, when Professor Emery Myers Emmert at the University of Kentucky, used the less expensive material in place of more expensive glass. Professor Emmert is considered the father of plastics in the U.S., because he developed many principles of plastic technology for agricultural purposes through his research on greenhouses, mulches and row covers.

EARLY DEVELOPMENT AND STATUS OF PROTECTED AGRICULTURE

Mulches. Natural mulches such as leaves, straw, sawdust, peat moss,and compost have been used for centuries to control weeds and hold moisture in the soil. None of these materials have been employed to any great extent in commercial vegetable production (Thompson and Kelly 1957).

It is only in the last fifty years that synthetic materials have altered the methods and benefits of mulching. Their potential for mulching was established through early research projects with polyethylene, foil, and paper.

Paper mulches attracted a good deal of attention in the early 1920's. They were not adapted for commercial vegetable production because of their short life, as well as the cost of material and labor, which was not mechanized (Hopen and Oebker 1976). In the late 1950's and early 1960's, improved formulations of paper - including combinations of paper and polyethylene, foils and waxes - stimulated research and the use of mulching materials. Petroleum and resin mulches for arid climates were developed at the same time. Of these mulches, only those made of polyethylene are still used today in the agricultural industry. The preferred colors are clear and black, although a wide variety of shades and colors are used for specific reasons in the production of food crops. Significant advances in the use of mulches occurred during the early 1960's with mechanization, the invention of mulch applicators and transplanters which would plant directly through the mulch.

Infrared transmitting (IRT) mulches, which transmit most of the solar heat portion of light radiation but absorb most of the visible portion, have recently been introduced to the market (Loy, et al. 1989). IRT mulches provide weed control as does black mulch but increase the soil temperature, as with clear plastic mulch. Unfortunately, labor requirements to remove plastic mulch from the field after the growing season can be high. New bio- and photo-degradable polyethylene and combinations of polyethylene-paper and polyethylene-starch show promise in eliminating the need for mulch removal.

Today, millions of hectares are planted to plastic mulch. In the People's Republic of China alone, over 2,867,000 ha. of mulch was used in 1989, a phenomenal increase over the 44 ha. in 1979. The area and tonnage used in each region of the world are listed in Table 1.

Table 1. Estimated world use of plastic mulch (1987-1989)

Region	Area (ha)		Tonnage (mt)	
	Min.	Max.	Min.	Max.
Western Europe	150,000	200,000	40,000	50,000
Eastern Europe	8,000	0,000	2,000	2,500
Africa and the Middle East	8,000	10,000	2,000	2,500
Americas	180,000	200,000	45,000	50,000
Asia and Oceania	3,000,000	3,329,000	330,000	366,190(*)
World Total	**3,346,000**	**3,749,000**	**419,000**	**471,190**

*In the People's Republic of China, consumption is only 0.1 mt/ha because the plastic mulch is a very thin film (8 MM) rather than the thicker film used in other regions.
Source: International Committee for Plastics in Agriculture (CIPA), 1989 (Anon., 1987) and Chinese Plastic Mulch Research Assoc. 1989 (Huang, 1989).

Row Covers. Row covers, or plastic tunnels, protect crops from frost and create favorable conditions for plants to achieve early production. Before the introduction of polyethylene, early spring crops such as cucumbers were started and grown in muslin-covered wooden box frames measuring approximately 17 meters square at 0.3 meters in height. This was a costly but effective method of producing early fruit from 1935 to 1945 (Hall 1963). In the mid-forties a method using two separate paper caps replaced the wooden box frame. A small cap, 28 cm. in diameter and 14 cm. in height, was used to start the plants. A second, larger, tent-type paper cover was installed when the plant filled the smaller cap. This second cap measured 35 cm x 28 cm x 21 cm in height. This tent cap was constructed so that one or both ends could be opened. Usually it was the leeward side which was opened and the plants were trained in that direction. The paper tent thus acted as a wind break. The early fruit could develop while the plant had partial protection during adverse weather. The double cap produced fruit as early as the wooded frame method but was less costly. Paper covers are still used today in some parts of the world. Paper covers have one serious liability: while they help protect plants from early spring frost and wind, they also reduce the amount of light reaching the plant, with the result that plants may be succulent and weak. In Japan, more translucent materials, such as vinyl or polyethylene film are substituting paper as plant covers or hotcaps. Such hotcaps not only protect against light frost but also provide extra heat and protection against chilling winds, blowing sand and soil particles.

Plastic row covers were initially used in Europe and the United States, and especially in Japan. In fact, in 1959, France and the U.S.A. totaled less than 400 ha. under plastic; Japan had more than 8,000 ha. (Buclon 1966). Since then, this method of protected agriculture has become common throughout the world.

Today, as in 1959, Japan uses mostly polyvinylchloride (PVC) film for row covers. In other countries, polyethylene predominates. There are historical as well as economic reasons for the selections of different materials for the same task. PVC films have a better heat-retaining (infrared radiation) capacity than polyethylene but they are also more expensive. Early in the development of row covers it was not possible to produce PVC sheets wider than 1.6 meters, although polyethylene films of 2-12 meters in width were available. With government financial support, Japan was the first country to develop wide PVC sheets (2-3 meters); as a result, Japan selected this material as the predominant type of film. France and Italy found the equipment for the extrusion-blowing process of making polyethylene much less capital intensive than the PVC equipment, and therefore selected polyethylene for use as row covers.

In the United States, the first use of polyethylene row covers for early crop production was for a cucumber planting in California in 1958. With careful venting adjustments for weather changes, plastic row covers produced a margin for marketing of four to five weeks over that of the two paper cover methods and produced good yields as well (Hall 1963).

For 25 years there was steady growth in the use of plastic row covers. However, no significant increase has occurred in recent years (Table 2), with the exception of the People's Republic of China where there are approximately 80,000 ha. under cover at the present time and expectations of greatly expanded use in the near future.

Throughout the world, a total of some 70,000 ha. are covered with PVC film: 60-62,000 ha. in Japan, 4,000 ha. in France and 1,500 ha. in Italy. Low density polyethylene film, on the other hand, is used on about 195,000 ha., of which over 80,000 ha. are located in the People's Republic of China (Huang 1989). In 1988, the hectareage increased over 30,000 ha. in China alone. The total world tonnage amounts to about 74,000-76,000 mt of polyethylene used for tunnels.

Table 2. Estimated world use of plastic row covers (1987-88)

Region	Area (ha)
Western Europe	60,000
Eastern Europe	20,000
Africa and the Middle East	10,000
Americas	10,000
Asia and Oceania	165,000
World Total	**265,000**

Source: International Committee for Plastics in Agriculture (CIPA), 1987 (Anon. 1987) and the Chinese Plastic Mulch Research Assoc., 1989 (Huang 1989).

The simplest and most economical form of row covers is the direct, or floating, covers with no sustaining wire or cane hoops. First introduced in Germany in 1970, floating covers then were adopted by neighboring countries. Perforated polyethylene film 50 mm thick, generally with 500 holes per m^2 (i.e. 4% ventilation, 46 g/m^2) now competes with non-woven/spunbonded fabric materials (PP, PA, polyester), which are porous and much lighter (10-25 g/m^2). These latter materials have been particularly successful in France (2,800 ha. out of 4,500 ha., and in Japan (100% of 4,000 ha.). According to the International Committee for Plastics in Agriculture (CIPA), the 1986 world estimates are:

- Perforated PE film: 15,000 ha. x 0.45 MT/ha. = 6,750 MT
- Non-woven veils: 10,000 ha. x 0.2 MT/ha. = 2,000 MT

Since the non-woven covers are quite new in agriculture, many grower and university experiments are currently underway throughout North America, Europe and Japan to understand the effectiveness of this new generation of plastics in modifying the micro-environment in crop production. For instance, scientists at the University of New Hampshire are studying the use of non-woven covers in production of a number of crops; including vegetables, small fruits, flowers, ornamentals, tree seedlings in nurseries, turf and the overwintering of vegetables, perennials and nursery stock (Wells and Loy 1985).

In the People's Republic of China, production of cucurbit crops under plastic row covers for early market has increased dramatically.

Greenhouses. The total world area of glasshouses over the last 10-15 years has been estimated at 30,000 ha. (Anon. 1987); with most of these found in northwestern Europe.

In contrast to glasshouses, plastic greenhouses have been readily adopted on all five continents, especially in the Mediterranean region, China and Japan. Most plastic greenhouses operate on a seasonal basis, rather than year round, as is the case with most glasshouses. The estimated area of plastic greenhouses is shown in Table 3.

Table 3. Estimated world use of plastic greenhouses (1987-1988)

Region	Area (ha)	
	Min.	Max.
Western Europe	55,000	58,000
Eastern Europe	16,000	18,000
Africa and the Middle East	15,000	17,000
Americas	8,000	10,000
Asia and Oceania	91,000	95,000
World Total	**185,000**	**198,000**

Source: International Committee for Plastics in Agric. (CIPA) 1986 (Anon., 1987), the Chinese Plastic Mulch Res. Assoc., 1989 (Huang, 1989) and Intn. Seminar on the Utilization of Plastics in Agric., 1988 (Park, 1988).

PVC film for greenhouses is still dominant in Asia, especially in Japan (35,200 ha), and low density polyethylene is also used in Italy (500 ha) and Greece. LDPE films cover a total of 149,000-162,000 ha; the average consumption is 1.5 MT/ha/year,

with a total world tonnage of about 224,000-245,000 MT/year.

In Japan, the area covered by plastic film greenhouses increased 35,000 ha. in just 20 years (1965-85). In Korea, these greenhouses increased 6.3 times, from 3,099 ha. in 1975 to 21,061 ha. in 1986. The People's Republic of China showed equally dramatic growth: 5,330 ha. in 1978 to 34,000 ha. in 1988. The combined growth for both greenhouses and row covers, in China exceeded 96,000 ha. in just ten years. Undoubtedly, China is one of the largest users of agricultural plastics in the world, where over one billion people - 20 percent of the world's population - are being fed from only 5 percent of the earth's cultivated land.

Since 1960, the greenhouse has evolved into more than a plant protector: it is now better understood as a system of Controlled Environment Agriculture (CEA), with precise control over air and root temperature, water, humidity, plant nutrition, carbon dioxide and even light. The greenhouses of today can best be seen as plant or vegetable factories. Almost every aspect of the production system is automated, with the artificial environment and growing system under nearly total computer control. In a research setting, such a totally enclosed system, with artificial light, is called a growth chamber or a phytotron. In the United States and Japan, such systems may cover large areas.

Controlled environment agriculture has gained in horticultural importance not only in vegetable and ornamental crop production but also in the production of plant seedlings, either from seed or through tissue culture procedures.

In the last 15 years there has been increasing interest in the use of soilless or hydroponic techniques for producing greenhouse horticultural crops. The future growth of greenhouse or controlled environment agriculture, where hydroponics is used for vegetable production, will depend greatly on the development of production systems that are competitive, in terms of costs, with open field agriculture.

PART TWO

PROTECTING MATERIALS AND STRUCTURES

CHAPTERS 3 - 4

3

COVERING MATERIALS

The financial return for producing early crops can be high, but, so is the risk. Protected agriculture can reduce this risk.

Off-season crop production means more than just early spring cropping; it includes production during times of adverse climatic or economic conditions, such as may occur in summer or winter. For example, off-season production of greenhouse melon crops in the United States is not profitable due to competition from melons grown in Mexico. The slightly warmer temperatures in Mexico obviate the need for protected agriculture, except perhaps for row covers, while in the United States the more expensive methods of greenhouse agriculture are still needed to produce a similar crop. This weather advantage - and lower labor costs - gives Mexico the economic advantage in producing off-season winter and early spring melon crops, despite the high costs of shipping the crop north.

Of prime importance in offsetting the additional cost of protected agriculture is a good yield. To this end, many methods have been devised to protect crops against the cold and frost. Ideally, agricultural production would be located in areas that are frost-free; however, in many countries, such locations are rare or even absent. In addition, increased capital and transportation costs, as well as excise duties on imports, often make it unprofitable for farmers to produce horticultural crops in another country for exportation back to the home country.

Farmers have devised many methods of protecting agriculture, with varying degrees of effectiveness. These include smoke, wind machines, wetting agents, chemical fogs, bacterial spraying, sprinkling, hot caps, and brushing. Smoke, which has been used by some vegetable growers in the United States to produce a cloud cover during cold periods, has come into conflict with environmental pollution laws and policies. Wind machines prevent frost by moving air over the endangered plants. In field trials, a wind machine in the 100 h.p. class has been shown to provide protection of 2°C or more for an area of 3-3.5 hectares on a clear, calm night. Such systems are commonly used in fruit orchards in the United States. Wetting agents and chemical fogs have demonstrated their effectiveness in protecting plants during experiments but are rarely used in commercial production because of cost and lack of protection during times of slight air movement. Bacteria sprayed onto plants give a small degree of frost protection; however, governmental agencies have been slow in granting permission to use such bio-controls because of uncertainty about the impact of these bacteria on the overall ecology.

The sprinkling of water - overhead irrigation - at the time of frost is often used to prevent damage but has serious disadvantages. Water turning to ice releases heat energy, therefore, the latent heat of freezing water keeps the plant tissue at 0°C; and even offers good protection to several degrees below 0°C. However, covering a large area at once with a sprinkler irrigation system can be prohibitively expensive because of the cost of pumping and sprinkler equipment. In addition, sprinkler irrigation usually wets the soil more than necessary; this results in the leaching of nutrients and a drop in soil temperature. As a consequence, plant development is retarded and the earliness of yield is affected. Overhead irrigation also requires careful monitoring of proper rates of water application to prevent damage from ice formation. This occurs when excessive water, with large droplets, accumulates, followed by a prolonged period of freezing.

Individual plant covers, or hot caps, made of paper have been used for many years to protect plants and hills of seeds or plants in the field, but are not used extensively today. In Michigan, beginning in the 1920's, covers made of paraffined or oiled brown paper made it possible to field plant approximately two weeks earlier than usual. Although these covers helped to protect plants from the early spring frosts and winds, it was found that the brown paper used, reduced the light to the plants and tended to make them more succulent and spindly in growth. Similar growth occurred when other materials were used, such as celluloid, glassine, paper parchment, and brown wrapping paper. After the paper covers were removed, the plants were easily damaged by a cold wind or a light frost because of their soft, succulent growth (Hibbard 1926).

Plant covers of vinyl or polyethylene film are still used in Japan. Framed with wire or split bamboo sticks, with certain crops, they produce an earlier marketable product with higher total yields.

The "brushing" methods of crop protection consists of leaning shields of brown kraft wrapping paper over the plants. These shields are attached to a framework of brush, lath, or wire and placed against supports on the north side of the east-west rows. With this, a micro-climate is formed and earliness of yield can be accomplished. A similar system, using palm fronds, is employed in the Middle East during the early fall months to protect horticultural crops from hot winds and direct sunlight. This method is not feasible for extensive areas since it requires large amounts of plant material, which is often not readily available. Such methods of crop protection are rarely used today, especially since the introduction of agricultural plastics.

While wind machines are quite common in fruit orchards, especially in citrus production, none of the above systems of protected agriculture is as common as mulching, row covers/low tunnels, and greenhouse agriculture.

MULCHES

Mulch systems are primarily used to increase soil temperature, reduce compaction, prevent weed growth and conserve soil moisture. All of these increase crop growth and production of saleable product and, in some cases, promote earliness.

Organic Mulches. These types of mulches are not used to increase the earliness of crop production, with the exception of petroleum mulches. Straw is sometimes placed over low-growing crops such as strawberries to give protection against freezing winter winds that may dessicate the strawberry plants, sometimes causing severe damage. In China, melon crops are commonly grown in the stubble of small grain crops. This placement permits fruit to grow and mature on a bed of straw mulch, which keeps it from contact with soil-borne fungus, lessening the incidence of damaging fungal diseases.

Petroleum Mulches. Since the early 1950's, petroleum mulches, sometimes called liquid mulches, have attracted a great deal of attention. These water emulsions of petroleum resins are formulated for spraying by adding an appropriate solvent and a suitable surfactant that is emulsified into water. The formulation produces a gel-like film that is sprayed onto the surface of the soil. Yield increases of as much as 50 percent have been reported with petroleum

In Kuwait, strawberries are grown in polyethylene greenhouses on plastic mulch, greatly reducing evaporation of drip irrigated water purified by desalination.

mulch on carrots, onions, lettuce, turnips and radishes (Milner 1963).

Most of the value received from petroleum mulch is derived in the first few weeks after planting, during the period of germination and early growth. Petroleum mulch is rarely used in agriculture because of its two strong disadvantages: it does not control weeds and it is a very messy material to mix and apply.

Plastic Mulches. Synthetic mulches such as those made of polyethylene are now common throughout the world. Plastic mulch is available in sheets of various widths, colors, and thicknesses. The width most commonly used is 1.3 meters, with a thickness of 37.5 microns. Black is the most common color of polyethylene mulch. However, there has been increasing interest in the use of clear plastic since its clarity increases the soil temperature, thereby promoting earlier growth and higher yield than black plastic.

Plastic mulch provides the following advantages (Lamont 1991):

1. Earlier crops. By raising the soil temperature in the planting bed, plant growth is accelerated, producing earlier yields. Black plastic mulch can result in 7-14 days earlier harvest while clear plastic accelerates harvest by 21 days in many conditions. Caution must be taken in using plastic mulch in hot desert regions, especially with crops such as peppers, that do not shade the mulch later into the summer. This results in excessive temperature build-up in the soil, which encourages certain fungal pathogens such as *Pythium* spp to thrive and often destroy the crop. White or white-on-black mulch creates a cooler soil temperature than either clear or black. This mulch is preferred for establishing crops such as fall tomatoes under hot summer conditions.

2. Reduced evaporation. Mulch reduces soil water loss. Because more uniform soil moisture is maintained, the frequency of irrigation may be reduced.

3. Fewer weed problems. Black and white-on-black plastic reduces or eliminates weed problems, as do other mulches which block the transmission of most of the photosynthetically active radiation.

Because clear plastic allows nearly full light transmission, weed growth can be a serious problem; for this reason, black is sometimes preferred. Application of herbicides to eliminate weed problems is becoming increasingly constrained by environmental concerns and legislation. Grey plastic mulch provides some advantage in weed control. The grey color transmits a limited amount of light but still enough to promote weed growth. Nevertheless, since much of the infrared radiation is stopped at the surface of this plastic film, the surface heat is sufficient to burn weeds when they touch the film. With less weed growth under the grey film there is less loss of fertilizer nutrients and less water loss from the soil. Since more infrared radiation permeates the grey plastic than the black, more heat reaches the soil, resulting in an earlier harvest.

4. Reduces nutrient leaching. Nitrogen and potassium are easily leached without plastic mulching.

5. Reduced soil compaction. Soil under the plastic mulch remains loose, friable, and well-aerated.

6. Root pruning eliminated. Cultivation is eliminated, except for the area between the mulch strips.

7. Cleaner vegetable product. The edible product does not come into direct contact with the soil, which results in a cleaner product with less rot and fruit blemishes. The cleaner product requires less attention in grading, packing, and processing.

8. Aids fumigation. Mulches increase the effectiveness of chemicals applied as soil fumigants.

9. Reduced water-logging of crop. Water is shed from the row area by the raised, tapered bed although not all plastic mulch is used only on raised beds.

10. Assist in insect management strategies. Use of reflective mulch helps to repel insect vectors of virus diseases.

In the mid-1980's, a plastic mulch called "Infrared Transmitting", or IRT mulch, was placed on the market. IRT occupies a niche between black and clear mulch, affording weed control as with black mulch and increasing soil temperature as with clear mulch (Loy and Wells 1989).

IRT absorbs (or blocks) most of the visible radiation, that part of the solar spectrum which supports photosynthesis and the growth of weeds. In areas where either purslane (*Portulaca oleracea L*) or grass species cause serious weed problems under clear plastic (without herbicide treatment) there is little or no weed growth under the IRT mulches. Weeds will germinate under IRT mulch and will either continue to grow slowly during cool weather or will be largely killed by high temperature under the mulch surface when daytime ambient temperatures are higher than approximately 27°C.

If properly managed, plastic mulch can produce significant yield increases. Plastic mulch has increased yields in several crops, such as tomatoes, peppers, eggplant, muskmelons, summer squash, cucumbers, watermelons, and strawberries, even doubling production in some cases (Geraldson 1962; Nettles 1963).

Carolus (1962) has demonstrated that a polyethylene mulch increased the early yield of planted tomatoes by 50 percent with a 300 percent increase in growth. In a cool season, muskmelon varieties increased in yield from 40 to 203 percent. The yield of early cucumbers was increased by 130 percent and the total yield by 30 percent. Woodbury (1963), in Idaho, compared black plastic mulch with non-mulch production and got a yield of 8,733 muskmelons on the black plastic treatments versus 2,994 melons without plastic. The mulch advanced peak production by 1-2 weeks.

Lamont (1991) lists possible yield achievements using plastic mulch systems with drip irrigation:

At the University of New Hampshire, research comparisons between IRT, clear and black plastic mulch showed that IRT doubled early melon yields over black mulch with over 33 per-cent more total yield (Loy 1991). Clear mulch produced nearly 13 percent more early yield than IRT but produced slightly lower total yield due to weed problems under the clear mulch. Melon plants growing on the IRT and clear mulch did not exhibit cold injury as did those on black mulch. In the first month, vine growth was double that obtained on vines growing on black mulch.

No doubt plastic mulch can increase yield and improve product quality by modifying soil temperature and controlling soil moisture. A mulch can facilitate fertilizer placement and reduce the loss of nutrients through leaching. Mulches can also provide a barrier to soil pathogens.

Reflective mulches have been shown to repel certain insects such as aphids (*Aphis gossypii*). The reduction of aphids and other vectors can greatly reduce the incidence of virus diseases. In tests conducted by Dudley et al. (1980) muskmelons were grown on four synthetic mulches: aluminium foil on paper, aluminum on black polyethylene, black polyethylene and white on black polyethylene. All mulches produced lower aphid counts than unmulched plots with the aluminum foil on paper having the lowest overall count. The soil temperatures were highest under the black mulch and lowest under the aluminum foil on paper. The largest early yield and the greatest total yield was from the black mulch treatment.

These trials were conducted in the spring when aphids numbers were low. During late summer when the aphid populations may be high in some areas, the incidence of virus may reduce yields by as much as one-half (Dudley et al. 1980). Using black mulch in the late summer, during a time of already high soil temperature, may heat the soil excessively, thereby damaging plant growth rather than promoting it, as happens during cooler periods of the year. Reflective mulches offer the advantage of not overheating the soil at a time when they are needed to repel insect vectors of severe virus diseases. Their use has therefore shown promise in tropical regions where insect populations are high year-around.

Early research results with colored mulches, testing the effect on plant growth, yield, and insect populations (Decoteau, et al. 1986; Kaplan 1991) look promising. While too early to recommend, this new development is the first major application to come from the scientific discovery of phytochrome.

Despite the obvious advantages, the high cost of synthetic mulches has limited their use in commercial production. At present, synthetic mulches are used for crops with high value per unit area. However, plastic mulch is increasingly used for the production of cotton in Israel and maize in France, as well as fruit trees and vine crops.

In China, plastic mulch aids in the production of forty different crops, includin cotton, corn, rice, transplants, fruit trees, tobacco, sugar cane, peanuts, and vegetables, and the Chinese have conducted research on the use of mulch with more than 90 different crops. The total area of

Fruit decay and blemishes are reduced when plastic mulch is used to prevent direct contact of fruit with the soil.

mulch in China has reached 2.87 million hectares, with 65 percent in the north and the rest in the south. Figure 1 shows the breakdown of crops grown on plastic mulch.

The cost of mulch has been reduced because the mulch used is a very thin film rather than the thicker film used in other countries.

The material being used for mulch film is mostly low density polyethylene (LDPE) but some LLDPE (extra low density) and high density polyethylene (HDPE) is also used. Generally, LDPE film is 0.014 mm; however, film made of LLDPE, HDPE and LDPE mixed with HDPE along with LLDPE mixed with HDPE is only 0.008 to 0.01 mm thick. The very thin films are very popular with the growers, since the growth benefits of the thin mulch are the same as the 0.014 mm ones and cost 30 percent less. Most of the mulch films are clear although some silver, black, and white films are used. Herbicide films and photodegradable films are being studied and tested.

Disposal. The disposal of waste mulch is of great concern as large landfills become overburdened with waste plastic. Polyethylene mulch does not decompose and must be removed from the field or it will interfere with future tillage.

In Japan, the handling of waste plastic is one of the biggest problems yet to be solved. Plastic consumption has increased dramatically in Japan in recent years (Takakura 1988). In 1985, the total amount of waste exceeded 165,892 tons, which included waste materials from greenhouses and row covers, as well as plastic mulches. Since 1970, Japan has treated plastic waste under the law of industrial wastes. Growers are themselves responsible for handling the wastes and, in the process, must not produce any air or water pollution. It is illegal to carelessly discard the waste plastic in a manner that might create obstacles in rivers and other public places.

The three methods used in Japan to discard plastic waste are: (l) recycling, (2) burial, and (3) incineration. Takakura (1988) explains the methods as follows:

Recycling. Five types of recycling are used. a) Generation of pellets and fluff. Collected waste plastics, mostly PVC, are first graded and foreign matter is removed. After rough crushing, they are washed with water, finely crushed and dried. The reproduction ratio of used materials is approximately 50% for PVC. The products are half-materials for plastic tiles, mats, sandals and fillers. b) Collected waste plastics, either PVC, or PE are crushed and then melted without washing. Plastic exudation makes final plastic products. c) Collected waste PE is crushed and mixed with sawdust or rice hulls to make solid fuels whose calorific values vary from 5,630 to 10,050 kcal/g, which are equivalent to those of coals and cokes. d) Waste PVC and PE can be treated to make hydrophobic materials for drainage. e) Oil or gas can be recycled by pyrolysis of wasted PE.

Burial. Waste plastics which are not suitable for reproduction must be buried according to the law regulating plastic waste disposal. The place and method of burial and pre-treatment are regulated by the law.

Incineration. Incineration is also regulated by the law. Incineration of amounts up to 100 kg/day by an

Figure 1. Crops grown on plastic mulch in China

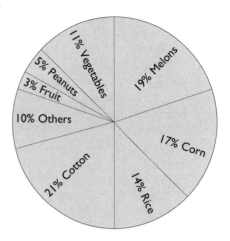

authorized system is allowed, although disposal of such large amount is not recommended. More detailed and thorough approaches are needed, both technically and administratively.

Recycling of plastics in Finland (Cornwell 1989) is a major business for a private company producing heavy-duty plastic sacks, agricultural films, and construction grade films. The company collects used films from the community and returns to the plant to process them. The film is separated by type, whether clear, colored, or printed; it is then washed, dried, and repelletized for feedback into the cycle. Since reprocessed resin is not of the quality of virgin resin, only 15 percent, or less, reprocessed plastic is used with virgin raw material. Except for medical or food packaging, injection molded plastic processors use half reprocessed plastic and half virgin material for products such as furniture and toys.

Degradable plastic mulches are currently receiving much attention, especially the photodegradable mulch. These plastic mulches have many attributes of standard polyethylene mulch: they are easy to lay and provide the usual benefits associated with mulch. The major difference is that photodegradable mulches decompose after the film has received a predetermined amount of UV light. The chemical composition of the film determines the amount of light required to initiate breakdown.

When the degradable mulch has received sufficient light it becomes brittle and develops cracks, tears, and holes. Small sections of mulch (usually less than 5-6 square cm) may be torn off and blown away by the wind. The film finally disintegrates into small flakes and disappears into the soil. The edges of the mulch covered by soil will retain their strength and decompose very slowly (Garrison 1990). To facilitate quicker breakdown of that mulch buried in the soil, soil covering the edges should be removed before final harvest, or as soon after harvest as possible. Exposing the covered edges to light will initiate the breakdown process, soon enough it is hoped, to facilitate sufficient decomposition before field preparation the following spring.

Each material responds differently to conditions within a given growing region. There are regional variations in light intensity, air and soil temperature, and soil type. Shading of the mulch and the amount of light affects the time of breakdown. Other factors that influence the time of breakdown are: plant growth habit (vine or upright), the time of year the film is laid (early spring or summer), the time between laying the plastic film and planting, the crop vigor, and the use of double or single rows (Garrison 1990). The most important factor affecting breakdown is the formulation of the mulch - such as whether it is a short, intermediate, or a long-lasting film. Because of these considerations, it is best to experiment with the various options and manufacturers of film before a specific photodegradable mulch is selected for large-scale application.

Biodegradable mulches are still in the experimental stage. Research only began in the mid-1970's (Otey and Westoff 1980). The most promising film formulations contain about 40 percent starch and 30 percent each of poly (ethylene-co-acrylic acid) and polyethylene. In research at the Argonne National Laboratory, wastes of potato starch and cheese whey are being put through a fermentation process; the end product, lactic acid, appears to be a viable candidate for conversion to environmentally safe degradable plastics (Bonsignore et al 1990).

When good biodegradable mulches come onto the market a great breakthrough will have been made in reducing the cost of plastic removal from the field and eliminating the problem of plastic disposal.

Synthetic mulches will continue to be used, particularly in the intensive cultivation of valuable crops. Certain conditions accelerate the spread of synthetic mulch culture, notably:
• water shortages in arid and semi-arid regions,
• the need for increased production, in limited space and over longer seasons,
• a shortage of plant nutrients, requiring more efficient fertilizer use,
• the need for non-chemical pest and weed control, and
• sanitation regulations for fresh produce which requirethe crop to be kept away from the soil surface.

Research has already demonstrated the degree of improvement that various mulches make on plant productivity. Future research will probably concentrate on understanding the microclimate provided by the different mulches, with the goal of determining how each microclimate can be regulated to provide optimum conditions for specific plant species.

The suggested plant spacing for vegetables on plastic mulch is listed in Table 4. Suggestions on double-cropping of vegetables on plastic mulch is listed in Table 5.

ROW COVERS
Since the mid-1950's, row covers have become an important method of protected agriculture. Early work in Japan (Shimokawa and Ono 1954) showed that the use of vinyl film tunnels permitted cucumbers to be planted in April instead of early May, advancing the harvesting by a period of 10 days and increasing the yield of 175 percent. Early work by Vogal (1963) found that the harvesting of spring crops of carrots, lettuce, cauliflower, kohl-rabi and rhubarb was advanced by an average of two to six weeks by covering the plants with plastic tunnels for a short period in the spring. With these particular crops he found that while row covers advanced harvesting, they had little effect on total yields.

Table 4. Suggested plant spacing for vegetables grown on plastic (Lamont 1991)

	In-row Spacing		Between row
	Single row	Double row	Spacing
	cm	cm	cm
Cucumbers (slicers)	30-45	22-45	30-35
Cucumbers (pickles	30-45	22-45	30-35
Eggplant	45-61	45-76	35-40
Honeydew	45-76		
Lettuce (leaf)		15-22	22-30
			(3 rows)
Muskmelon	45-76		
Okra	30-45	45	35-40
Pepper	30	22-30	30-45
Pumpkin	61-121		
Squash (summer)	30-40	40-61	5-40
Squash (winter)	45-121		
Tomato	45-61		
Watermelon	61-121		
Broccoli		20-30	22-30
Cabbage		22-30	30-40
Cauliflower	45	45-61	35-45
Chinese cabbage	30	22-30	30-35
Collard	22-30	30-45	30-45
Corn (sweet)	15	15-30	30-45
Greens		15-30	22-30
			(2-3 rows)
Onion		10-15	10-25
			(3-6 rows)

Table 5. Suggestions for double-cropping vegetables using plastic mulches (Lamont 1991)

Crop Planted lst	Crop Planted 2nd
Peppers	Summer squash, cucumbers, or cole crops
Tomatoes	Cucumbers, summer squash or cole crops
Summer squash	Pumpkins, tomatoes, or cole crops
Eggplant	Summer squash
Cucumbers	Tomatoes, pumpkins, or summer squash
Muskmelons	Tomatoes
Watermelons	Tomatoes
Honeydews	Tomatoes
Broccoli	Summer squash, pumpkins, muskmelons, tomatoes
Cabbage	Summer squash, pumpkins, muskmelons, tomatoes
Cauliflower	Summer squash, pumpkins, muskmelons, tomatoes
Lettuce	Summer squash, pumpkins, muskmelons, tomatoes
Snap Beans	Summer squash, pumpkins, muskmelons, tomatoes
Sweet corn	Summer squash, tomatoes, okra or cucumbers
Onions	Tomatoes, snap beans or cucumbers
Herbs	Broccoli, cabbage, cauliflower, Chinese cabbage
Strawberries	Tomatoes, summer squash, cucumbers, muskmelon, pumpkins, okra

There are many different methods of using row covers. The following are systems that have proven viable for those regions of the world where row covers are currently in use.

California System. In California (Roche 1964) several methods are used to form the plastic tunnels over rows of food crops. For cucumber production, two sheets of 90 cm polyethylene film are formed into the sides of the tunnel. Growers in San Diego, California tested various types of hoops (Hall and Besemer 1972), deciding on 9-gauge, galvanized wire, cut in 175 cm lengths to support the two sheets of polyethylene in the formation of tunnels. The wire is bent into an oval shape (or hoop) spanning 70 to 80 cm., with the height of hoop established at 37.5 to 40 cm. The hoops are spaced 1.66 to 2.3 m apart. The row covers are reinforced by 2.5 x 2.5 x 70 cm wooden stakes driven into the ground below the center of the wire hoops at spacings of 3.5 to 5 m down the plant row. The hoops, in turn, are fastened to the stakes and wire to give stability in strong winds. Soil is used to hold the bottom edges of each polyethylene sheet. A 16-gauge wire stapled to the top of the stakes is the connecting fulcrum for the two plastic sheets. Ordinary spring grip clothespins secure the two plastic sheets to the top wire. In areas of high winds, a second wire hoop is doubled over the top of the plastic at every second or third hoop. This is especially important when the plastic covers are pulled back during times of ventilation. When ventilation is desired, the plastic is simply slid down one side of the tube between the wire hoops; and it is either fastened there by the clothespins or held in place by the combined force of the hoops. The top hoop secures the plastic between the hoops and reduces the flapping that may loosen and damage it. The film used is 38 microns in thickness and is clear.

For tomato production, the plastic tunnels are formed in the same manner as for cucumber production except that the wooden stakes are two meters in length, placed l-l.2 m apart down the plant row. Two wires are stapled on alternate sides of each stake at a height of 50-55 cm from the ground. Wire hoops are set at alternate stakes in forming the tunnel shape. Clothespins are used to hold the two 1 meter sheets to the two top wires.

Venting is one of the most critical features in using plastic row covers. In venting the above mentioned tunnels, the plastic is opened as needed and secured to the wire hoops and top wires with clothespins.

In Southern California, all of the early cucumber and tomato plantings started in January and early February are grown under row covers. Cucumbers planted as transplants in early February normally begin producing in mid-April, continuing

The "California System" of row covers using perforated plastic are common for growing early market tomatoes. As shown, the covers are closed during cool periods.

During warm days, the covers are opened for ventilation. As plants progress into the warmer weather, covers are left open and vines trained to wooden stakes.

tom to allow normal pollination. Clear plastic mulch is nearly always used in combination with the row covers. Mulch under a plastic tunnel will increase the day and night temperatures of the soil, at a depth of 5 cm, by 7-l0°C and 2-5°C respectively (Tarakanov and Rozov 1962).

The Fernhurst System. This system was first developed for strawberry production but is used throughout Europe and the Middle East for other crops as well, such as melons, cucumbers, peppers, and eggplant.

The polyethylene sheet, 130 cm in width and 38 micron in thickness, is spread over wire hoops that are placed at intervals of 60-90 cm in the plant row. It is not advisable to increase the distance between hoops since snow loading may distort or flatten the tunnels, permitting winds to remove the plastic cover. The hoops, 180 cm in length, are made of 8 gauge galvanized wire. Each hoop has two eyes made in the wire by a simple jig made on the farm. Each eye has a diameter of approximately 1.8 cm and the length of each "leg" is 20 cm.

Once the polyethylene is put over the hoops, the end of each row has an additional hoop for extra stability and the end of the poly sheeting is buried in a trench.

Lengths of polypropylene baler twine are cut 150 cm long, and loops are tied at each end. These are slipped over the eyes of the hoops to hold the sheeting in place. After securing the sheeting, the eyes of the hoops are pushed into the ground so that there is no gap between the polyethylene and the soil (Figure 3).

Ventilation of tunnels is easily achieved by lifting one edge of the film away from the soil level and pushing it towards the top of the hoop. It moves easily between the wire and the twine, but because it is securely tensioned at all times, it remains firmly in position even when the tunnels are opened (Figure 4)

through June. Tomatoes transplanted in late January will be ready for the first harvest in late May and continue to produce through mid-July.

Clear plastic mulch is used in combination with the row covers, where weed control is economically accomplished through the use of soil fumigation. The cost and cultural operations associated with clear plastic have been a limiting factor. Fumigation with methyl bromide and chloropicrin has given good weed, disease, and pest control (Hall and Besemer 1972).

Today, new food and drug regulations in the U.S.A. prohibit the use of many chemical fumigants, therefore alternatives such as soil solarization are employed. In soil solarization, clear plastic tarps or mulch are placed over the soil surface for a period up to 90 days during a summer fallow period to allow the soil temperature to rise approximately l0°C above bare soil at a depth of 5 cm. Such solarization procedures have significantly reduced the incidence of disease (Katan 1981).

In another row cover method in California, a single 150 cm sheet of clear plastic is centered over tall stakes and forced downward. The stakes poke holes through the plastic, which rides down to about the 50 cm level, where a lengthwise wire forms a tent ridge. Wire hoops often are used between the tomato stakes to form rounded tunnels, giving the plants more air space. This type of row cover is often used for rain protection.

Row covers are not commonly used in strawberry production in California, except for early January through April production when early production brings higher prices. Row covers also provide important rain protection. The cover consists of one sheet that is secured between two galvanized wire hoops with the aid of clothespins. For strawberries, the tunnel is open at the bot-

The "Fernhurst System" is commonly used throughout the world for crops not commonly trained upright to a trellis system of wooden stakes and string.

Figure 2. The Fernhurst hoop system

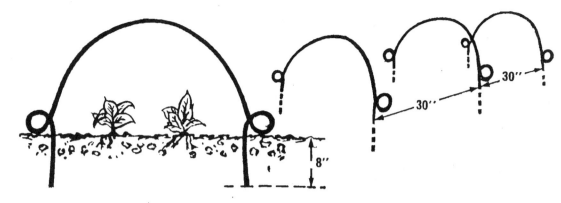

The hoops are spread to a width of 60 cm over the crop, ready for covering with polyethylene. The hoops are spaced 75 cm (30") apart with each "leg" pushed into the soil to a depth of 20 cm (8").

Figure 3. Side view of tunnel

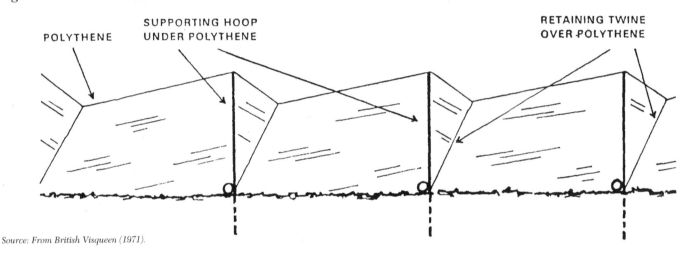

Source: From British Visqueen (1971).

Securing the polyethylene (Polythene) sheeting is essential in order to avoid any possible displacement by the wind. The securing twine must be fixed to the eyes in the hoops.

Figure 4. Ventilation of the tunnel

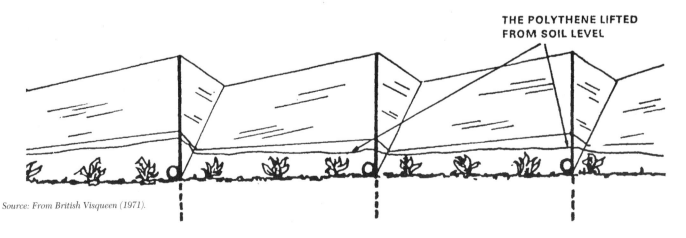

Source: From British Visqueen (1971).

Ventilation can either be provided at intervals along the row, or the whole of one edge can be lifted a few inches above soil level.

Figure 5. Wire hoops and ventilation slits

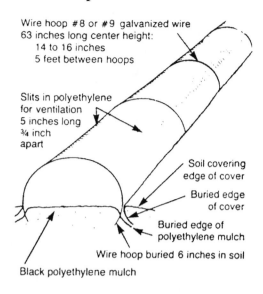

The wire hoops are 158 cm (63 in.) long, installed at a height of 35-40 cm (14-16 in.) in the plant row. The slits for ventilation are 12.7 (5 in.) long and 1.9 cm (.75 in) apart.

For spraying against pests and diseases, the tunnels are opened by lifting both edges of the sheeting to the tops of the hoops. To prevent the sheeting from slipping, it can be held in position by temporarily releasing some of the lengths of polypropylene twine, which are wound once around the bunched film and resecured to the eye of the hoop.

At blossom time, the tunnels are ventilated on one side by lifting an edge of the sheeting about 30 cm every fifth hoop. In England, irrigation by overhead spray line has been used to protect crops under the polyethylene from frost.

The twine is used instead of an additional wire that was once commonly placed over the top of the tunnels in order to secure them during wind conditions. The polypropylene twine is much lower in cost and does not cut or damage the polyethylene as wire might.

For general guidance, the wider the tunnel the better will be the temperature conditions prevailing in the early part of the year. However, as the width of the tunnel increases, so does the likelihood of wind damage. For certain low crops, particularly radish and lettuce, the base hoops can be spread out to give a flatter and lower tunnel. A low tunnel offers less wind resistance. Tender crops can be started very early under tunnels and part of the polyethylene can be cut away later to allow the crop to emerge. This technique is particularly suitable for runner beans, squash and sweet corn.

Perforated Plastic Tunnels. High winds remain a problem with every method of ventilation tried. A technique has been developed in France whereby the plastic is perforated in order to facilitate ventilation (MaMaire, 1964), instead of the conventional method of opening and closing the tunnels daily.

There is a slight difference between temperatures - minimum and maximum - with the perforated and the non-perforated tunnel. They produce similar results in earliness of harvest, and yield. From a practical point of view, the perforated

plastic tunnels are less labor intensive: solid plastic tunnels require considerable time to secure ample ventilation in the morning and evening during an eight week period. Little condensation occurs under the perforated film. Ettinger (1964), in Israel, demonstrated that perforated plastic sheets reduced tunnel temperatures by 5-6°C in comparison to unventilated solid plastic tunnels and that the perforation in the plastic reduced the need for manual ventilation.

Perforation can be produced by drilling the polyethylene, while it is still on the roll, with a low speed twist drill, aiming the drill at the center of the core. Excessively rapid drilling melts and fuses the film. The perforations are approximately 0.625 cm in diameter, spaced 8cm on center in each direction.

Slitted Row Covers. Slitted row covers have a series of crosswise slits which provide ventilation on sunny days. Without these slits, plastic covers would have to be manually opened and closed to provide ventilation during the day and cold protection at night. This method of ventilation was first developed at the University of New Hampshire in the early 1970's (Wells and Loy 1985). Using research on muskmelons, they showed that melons with covers plus black polyethylene mulch produced yields three times that of bare soil, having no mulch or row cover, and about two times greater than with black mulch alone (Loy and Wells 1974). Row covers and mulch together increased fruit maturity by 3-9 days with commercial hybrids and 12-13 days with experimental hybrids over those grown on black polyethylene alone.

In New Hampshire, row covers were initially patterned after the California system; however, the two piece construction was found to be very laborious and also subject to severe damage by wind gusts (Wells and Loy 1985).

While perforated row cover provided reasonably effective ventilation without appreciably sacrificing temperature increases, (Dubois 1978) perforated row covers were not put on the commercial market in the United States as they were in Europe and the Middle East. To achieve ventilation, row covers in New Hampshire were constructed from a single sheet of plastic, 1.5 m wide, 38 micron thick, with two rows of continuous slits, 1.9 cm apart and 12.7 cm long. Figure 5 shows the design of wire hoops together with installation instructions for the slitted row covers.

In the New Hampshire trials, the slitted, one-piece row covers showed about an 80% reduction in installation labor, were self-ventilating, eliminating daily manual opening and closing of the covers and were able to withstand very gusty winds (Wells et. al. 1977).

In addition to melons, other crops such as cucumbers, tomatoes and peppers are grown with the slitted plastic mulch. Since the weeds were controlled through the use of black plastic mulch, the slitted row covers were care-free from the time of installation until the time of removal, 3-6 weeks later, depending on the crop and the weather. When ambient temperature reaches the range of 30-32°C, either the covers are removed or extra slits are made for tomatoes and peppers. For vine crops, the covers are left on until the appearance of the first female blossoms or until the vines reach the edge of the covers (Wells and Loy 1985).

Frost protection with the slitted covers is similar to that with

Figure 6. Structure for air-supported row covers

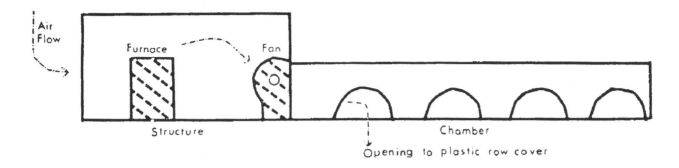

the perforated polyethylene covers but is inferior to solid covers. The maximum increase in temperature is only 1.0 to 2.0°C above open field temperatures whereas with solid covers, frost protection of 2.5° to 4°C can be achieved. In view of the advantages of the slitted tunnels over the solid, the use of slitted row covers is a reasonable compromise between maximum frost protection and a saving in labor. Row covers should not be viewed merely as a frost protection system but as a growth-intensifying system during cool spring weather. Therefore, it is not advisable to plant very early, hoping to protect these early plantings against heavy frost. An earlier planting date of 10 days to two weeks would be more reasonable (Wells and Loy 1980).

According to Wells and Loy (1980), late spring frosts are probably not the biggest restraint to early culture of muskmelons and other vine crops. Cool spring soils markedly inhibit growth, water, and nutrient uptake by young seedlings or transplants. Row covers, in combination with black plastic mulch, will usually increase soil temperatures by 3° to 4.5°C. Unfortunately, row covers will not protect vine crops, especially melon crops, against the effects of extended periods of five or more days of cloudy, cool weather. With such weather patterns, the soil temperature drops, thereby preventing the uptake of water by the plants. Consequently, when sunny weather does return, the young transplants wither as transpiration exceeds water uptake.

Air-Supported Row Covers. In an attempt to achieve a greater degree of cold protection, Jensen and Sheldrake (1965) applied artificial heat in plastic tunnels in order to provide frost protection to crops growing in the more northern latitudes of the United States (Figure 6).

Tomatoes, cucumbers and muskmelons were grown under row covers supported by air pressure from a fan located at one end of the tunnels. The speed of air movement through the tunnels was regulated by adjusting a small door opening at the opposite end of the tunnel. During periods of hot weather, ventilation or air movement through the tunnel was increased by fan speed and/or by enlarging the opening at the end of the tunnel. During times of frost, heat was added to the air stream through the tunnels. The smaller the door opening at the end of the tunnel, the less the air movement through the tunnels, causing a temperature build-up inside the tunnels during warm, sunny days. The larger the opening at the end of the tube, the faster the air movement or circulation, thereby lowering the temperature inside the tube to approximately that of the out-

side temperature. On one occasion the outdoor day temperature reached a high of 36oC. By fully opening the ventilation doors, the temperature at the end of the tunnels (50 m in length) reached a high of only 40oC. Temperatures of approximately the same degree were maintained in the wire-supported covers, which had open slits at the top of the tunnel for ventilation. Early in the tomato trials a low outdoor night temperature of minus 3oC was recorded. Those plants under the wire-supported row covers, with no added heat, were completely destroyed by the freeze. Those plants under the air-supported tunnels were saved with the application of heat.

When heat was supplied to the tubes the temperature decreased with the length of the tube. It was most important to have a small opening at the end of the air-supported tube; this opening allowed some air movement in the row cover, ensuring that the heat entering the tunnels would be distributed to the end.

Insect and disease control was accomplished by injecting a dust into the fan intake, which produced an even distribution throughout the planting. If the air supported row covers were kept rigid, they withstood the weight of snow better than those supported by wire; however any snowfall over 5 cm, especially wet snow, would collapse the plastic tunnels, thereby causing plant breakage. It was also important to keep the covers rigid during high winds so that they would not sway and injure the plants under cover.

Typically, muskmelons will mature 9-13 days earlier with wire-supported row covers in combination with plastic mulch than with black plastic alone. In the air-supported trials, due to the earlier than normal planting under covers supported with wire, the muskmelon harvest was 20-25 days earlier than the normal first harvest date of muskmelons not raised under covers. Tomatoes were harvested 15-20 days earlier than the normal out-of-door planting.

While there is usually no significant difference in early tomato production between plastic mulch and no mulch, there is normally a notable difference in the earliness of cucumber and muskmelon production. In trials conducted by Jensen and Sheldrake (1967) where air supported row covers were used in combination with plastic mulch, cucumber and muskmelon crops grown on plastic mulch produced much higher early yields than those with no mulch (Tables 6 and 7). Today, if growers are going to the added expense and effort of using row covers in combination with plastic mulch, they should consider using the IRT mulch.

Table 6. Effect of plastic mulch on marketable yields of early cucumbers, USA, summer 1965

Treatment	Cucumber var. Triumph	
	(No./ha)	Yield (kg/ha)
Clear plastic mulch	161,637	37,569
Black plastic mulch	148,592	33,336
No plastic mulch	115,507	26,043

Plant population: 13,610 ha.
Fruit harvest: June 29 - August 6
Source: Jensen and Sheldrake, 1967

Table 7. Effect of plastic mulch on marketable yields of early muskmelons, USA, summer 1965

Treatment	Muskmelon var. Harper Hybrid	
	(No./ha)	Yield (kg/ha)
Clear plastic mulch	13,230	13,362
Black plastic mulch	8,315	9,786
No plastic mulch	755	911

Plant population: 6,805/ha.
Fruit harvest: July 19 to August 16
Source: Jensen and Sheldrake, 1967

Again, row covers are used in combination with plastic mulch only when the goal is to gain an advantage in earlier fruit production over crops planted only through plastic mulch. While row covers will increase both labor and expense, they will raise both the air and soil temperature, provide some frost protection, and thereby increase earliness and yield.

While earlier yields of up to two weeks can be achieved in tunnels where heat is applied, the capital costs in fan and heating equipment can be quite high. Operating costs for electricity and propane can be expensive as well. The plastic row cover and mulching material may cost up to $1,750-2,500/ha.; the capital cost of the fan and furnace is approximately $1,500-2,000/ha.; and energy costs for fuel and electricity may reach $2,000-2,500/ha. These added expenses must be offset by higher returns in the market. Unless they are, then producing a product earlier does not justify the added cost of air-supported row covers. This was the experience in New Jersey where commercial use of air-supported row covers was discontinued after the first year due to the high cost of operation and maintenance.

Plastic Covered Trench System. In 1967, Garrison (1973) of Rutgers University, New Jersey, USA, developed a system of crop protection that permitted early seeding, promoted early plant development, and provided frost protection.

Polyethylene is stretched over a trench in the ground. The bottom of the trench is approximately 18 cm below the plastic. In the second method, plastic is laid across two soil ridges 40 cm in height. A mechanical mulch layer is used when the polyethylene was applied over trenches.

Weed control is essential under such systems of protected agriculture, and is normally accomplished through the use of chemical herbicides. The polyethylene is removed from the trenches when the plants first touch the plastic. With tomatoes, the trenches are filled and the beds leveled by cultivation within two weeks after the removal of the plastic.

Tomatoes produced under this method would come into production approximately 13 days before those seeded in the open field.

A similar system is presently used in California and Arizona to grow early cantaloupe and watermelons. In regions where there is geothermal activity, frost protection is effected by application of warm water through drip irrigation lines installed under the polyethylene. This system has saved melon plantings down to a temperature of -4.5°C.

A plastic-covered trench system is feasible where low temperatures limit early growth, where the danger of frost increases the risk of early planting, and where higher prices are received for early production (Garrison 1973).

While plastic covered trenches may not give the added growth response and earliness of yield in comparison to that of row covers used in combination with plastic mulches, this system is far less expensive and becoming quite common in the desert regions of the United States, Mexico and Israel.

Floating Row Covers. Floating row covers can offer protection to both cool and warm season crops. The simplest form of row cover is the fabric or floating row cover, without wire or cane hoops. First introduced in Germany in 1970, the technology was soon adopted by neighboring countries. By 1987, more than 15,000 hectares of floating covers were in use, mostly in Austria, England, France, Germany, Spain, and Italy (Hoag 1988). Their potential in North Africa and the Eastern

Muskmelons, in Ithaca, New York, transplanted directly to the field on May 5, as they appeared on June 5.

Mediterranean is being examined. They were first used in the United States in 1980, by Wells and Loy at the University of New Hampshire. Current use in the United States is estimated at 4,000 ha. (Mansour 1991). Floating row covers are made of spunbonded or non-woven fabrics: polypropylene (PP), polyamide (PA) or polyester. Polypropylene and polyester are the two fabrics most commonly available. These covers are made by melting the appropriate plastic, or combination of plastics. They are sprayed as fine filaments onto a moving belt which conveys them to a bonding roller. The roller presses and fuses the filaments together. This process is rapid and creates fabrics that are strong, lightweight, economical, and porous, ranging in weight from 10-50 g/m^2. Fabric durability is dependent on its weight, the type of plastic, the additives and the method of bonding. Fabric can be made into very wide pieces, ranging from one meter to about 10 meters in width; narrower widths may reach a length of 850 meters. By special order, fabrics of over 20 meters in width can be accommodated.

The covers can be applied over a single row (Figure 7) either by hand or by using a modified mulch applicator (Wells and Loy 1985), or over a number of plant rows with one large cover. When covering a single row, the material must not be stretched tightly but left with slack in the center to allow for expansion as the crop develops. The edges should be securely buried, especially when using lightweight covers, which may require weights, in addition to soil, to secure them. The wind will blow through the material. Using a combination mechanical/manual method of applying the widest covers to a field, a team of three people can cover nearly a half hectare in 40 minutes.

Heavier fabrics can be reused one or more times. Even the 17 g/sq.m material can be reused once if handled properly, if it

In the southwestern part of the United States, plastic covered trenches are becoming increasingly popular for the production of muskmelons for early market.

is left on a crop for only a short period of five weeks or less. The lightest weight fabrics are seldom reused (Mansour 1991).

The lightest covers, those around 10 g./sq.m are used as insect barriers. They offer protection against viruses and feeding damage from insects such as aphids, loopers, and beetles (Mansour 1991, Wheatley 1991). They also prevent or discourage feeding by birds and small animals. However, these fabrics are easily damaged by livestock, dogs, and other animals. They have minimal effect on temperature and light transmission.

According to Mansour (1991), those covers of about 17 g/sq.m are, by far, the most commonly available. They have the same applications as supported row tunnels or unheated greenhouses: All these protected systems of agriculture are used to enhance early maturity, increase early yield and total yields, improve quality and to extend the growing season or make possible the growing of crops in areas where they are not commonly grown.

Heavier covers, those greater than 30 g/sq. m, are used primarily for frost and freeze protection, and in situations requiring extra mechanical strength and durability to extend the growing season, and reuse of materials. Some manufacturers claim that the heavier fabrics will give frost protection down to -3.3oC. Polyester provides 1-2oC frost protection; polypropylene provides 2-3oC protection (Wells and Loy 1985). Some plant injury may result from the use of spunbonded materials as floating covers. As the plants mature, some plant foliage comes in direct contact with the fabric. Moisture forms more readily on leaf surfaces in contact with the covers; the heat that is readily reradiated from the surface of the row cover to the atmosphere causes more rapid frost formation on the leaf surface (Wells and Loy 1985). The heavier

JUNE 8

CLEAR PLASTIC MULCH

Muskmelons were far advanced when planted through a clear polyethylene mulch as compared to those planted without mulch.

Figure 7. Single plant row covered with floating row cover

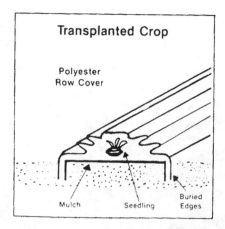
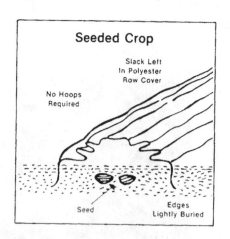

1. They modify temperature, increase heat units by 2-fold or more (Hemphill et al 1986a, Table 8), and they protect plants from light frosts. They help in the conversion of light waves to heat energy and trap this heat in the vicinity of the plants. Although permeable to vapor, under certain conditions of temperature and humidity, moisture can condense on the inside of these row covers. This affects air temperature as vapors condense to liquids on the covers, and release heat. The water droplets on the inside of the covers are also good heat sinks. They store this heat and radiate it to the crop at night. Condensation is a factor only under certain conditions when humidity under the cover is high and temperature differences between the air under the cover is higher than the surface of the cover.

materials reduce light transmission, and are therefore not used in the same way as the 17 gram materials (Mansour 1991).

Mansour (1991) reported the following in regard to floating row covers:

> Similar to row tunnels and greenhouses, floating row covers in the 17 gram range modify the plant microclimate in four major ways:

Table 8. Effect of insecticide sprays and floating row covers on yield and foliar virus symptoms, potato row cover trial, 1986 (Hemphill, et al. 1987)

Treatment	No. of virus plants/ plot[z]	Senescence on 8/8/86[y]	No. tubes per plot	Total yield (tons/ ha)	Mean tuber wt.(g)	Percent tubers with PVY Blossom end	Stem end
Main effects							
No spray	0.47	3.5	78.4	15.4	193.5	12.6	13.0
Sprayed	0.17	2.5	84.6	18.8	217.5	12.6	11.3
LSD (0.05)	0.30	0.4	NS	1.6	16.4	NS	NS
Reemay	0.00	3.0	71.3	15.8	219.6	4.0	4.0
Agronet	0.18	3.3	80.2	17.2	212.3	0.0	2.0
Agryl	0.08	3.0	79.4	14.8	181.8	0.0	3.3
No cover	1.00	2.8	95.1	20.5	208.4	42.7	39.3
LSD (0.05)	0.53	NS	12.1	2.3	23.1	9.3	11.6
Interaction							
No spray:							
Reemay	0.00	3.8	68.2	13.7	203.1	4.0	6.7
Agronet	0.20	3.7	76.3	15.2	199.6	0.0	2.7
Agryl	0.00	4.0	78.0	12.8	161.6	0.0	0.0
No Cover	1.67	2.5	91.0	19.6	209.6	46.7	42.7
Sprayed:	0.00	2.2	74.5	17.9	236.2	1.3	1.3
Agronet	0.17	2.8	84.0	19.2	225.0	2.7	1.3
Agryl	0.17	2.0	80.8	16.8	202.1	8.0	6.7
No cover	0.33	3.2	99.2	21.4	206.8	38.7	36.0
LSD (0.05)	0.75	0.7	17.2	3.3	32.7	13.1	16.4

[z]virus symptomatic plants from field observation of foliage.
[y]point scale, with 5 = foliage complete senesced, 1 = healthy greenfoliage

Table 9. The effect of slitted polyethylene and polyester row covers on early and total yields (kg/ha) of 'Goldstar' muskmelon over a 4-year period

		Test year			
Treatment	1981	1982[z]	1983	1984	Avg.
Early yield					
Black poly mulch	3466a[y]	0a	370a	672a	1127
Polyester & mulch	20,929b	1453b	17,473b	6451b	11,577
Slitted & mulch	21,377b	2560c	15,681b	6182b	11,450
Total yield					
Black poly mulch	50,099a	4574a	40,024a	27,507a	30,551
Polyester & mulch	67,433b	9453b	40,322a	29,478a	36,672
Slitted & mulch	65,978b	13,377b	52,420b	21,504a	38,320

[z]*Yield reduction in all treatments due to fusarium wilt.*
[y]*Mean separation within columns by Duncan's multiple range test, 5% level.*
Source: Wells and Loy 1985

2. The covers modify moisture by raising the relative humidity, and reducing water losses by transpiration or evaporation. Thus, they enhance seed germination by maintaining higher soil moisture at the surface, and prevent crusting.

3. The covers can modify gas concentrations around the plant. They may increase CO_2 levels when decomposing organic matter is available in the soil.

4. They modify light, thereby influencing photosynthesis, flowering and plant growth (Friend 1990).

Spunbonded field covers do the above things at a much lower cost for labor and installation than row tunnels or plastic greenhouses, thereby providing flexibility to the annual row crops grower who may choose to grow a protected crop one year but not the next.

Research at Oregon State University (Hemphill et al, 1987) has indicated that floating covers successfully exclude virus transmitting insects from potatoes (Table 8), and other research has indicated excellent protection from cabbage loopers. In the southwestern United States, growers use covers as insect barriers to protect vine crops from virus diseases that are difficult to control, such as zucchini yellow mosaic virus.

Before floating row covers are used in a region with no history in such methods of protected agriculture, the system should be tested first on a trial basis to screen local varieties of crops for their response to the covers.

In self-pollinating crops and leafy vegetables, the covers can be left on for almost the entire duration of the crop production period. Care must be taken to remove the covers during days when intolerably high temperatures might occur under the covers. Some cucurbits, such as muskmelon and watermelon, can tolerate temperatures well above 30oC and are thus well-adapted to row cover culture. Tomato and pepper foliage is reasonably tolerant of high temperatures, but fruit set under covers is impaired at temperatures over 30oC. These crops should not be grown under wide covers since the action of high winds on the covers will cause abrasion to the growing points of peppers and tomatoes. With spinach and lettuce, high temperatures can cause bolting.

Wells and Loy (1985) conducted research on floating covers over a four year period. They found that muskmelons, a crop sensitive to low soil and air temperatures but tolerant to high temperatures, responds well to row cover culture (Table 9).

Their research also illustrated yield increases with pepper (Table 10). Tomato yield increases were not as dramatic, especially, during one year when tomato yields under covers were drastically reduced due to high ambient temperature conditions.

Other crops, when grown under covers, have shown yield increases as well, such as sweet corn, edible pod peas, carrots, cabbage, leafy lettuce, green beans, squash, cucumbers, watermelon and root crops such as radishes (Table 11). The covers have proven to be especially effective in controlling cabbage maggots on radish and in reducing flea beetle damage, as shown in Table 11.

In the United States, the usefulness of floating covers varies with location and season. In the northern latitudes the covers are used in the early spring to extend the growing season by improving early yields as well as quality. In Florida, covers are used in winter production of tender warm season crops or the fruiting of crops such as strawberries. At times, they are used to protect crops from insect vectors.

The greatest benefits from the use of floating covers are realized when row covers and field covers are integrated into a total production system. This includes the planting schedule, proper management of the covers with regards to application, the timing in removal , and the management and use of plastic mulch, as well as irrigation and weed control under the row covers as well as the other benefits of mulch. Daytime soil temperatures are slightly higher with row covers plus black plastic mulch than with mulch alone, whereas nighttime soil temperates are only slightly elevated under slitted mulch and not significantly affected by polyester covers (Loy and Wells 1982).

In the future, new technology may provide row cover polymers which would change properties according to temperature. An ideal row cover (Wells and Loy 1985) would provide insulation at low temperature but would become more porous or more opaque at high temperatures in order to prevent excessive heat buildup. Materials used for row covers at the present time

Table 10. The effect of slitted plastic and polyester row covers on yield (kg/ha) of 'Greenboy' pepper

| | Yield | | | |
| | 1982 | | 1983 | |
Treatment	Early[z]	Total	Early[z]	Total
Bare soil	1613a[y]	5618a	1478a	9704a
Black plastic mulch	5887b	22,499b	1935a	19,864b
Polyester row cover[x]	12,876c	22,364b	2607b	12,741a
Slitted plastic row covers[x]	13,763c	24,111b	2124a	11,397a

[z]Early yield was first 3 harvests.
[y]Mean separation within columns by Duncan's multiple range test, 5% level
[x]Black plastic mulch was used with the row cover.
Source: Wells and Loy 1985

Table 11. The effect of spunbonded row covers on the yield (kg/ha) and insect control of 'Cherry Belle' radish

Treatment	No. of roots/ha	kg/ha	Maggot damage (%)	Flea beetle damage (%)
No cover	29,515[z]	731a	70	100
Row cover				
Polypropylene	114,405b	2000b	0	0
Polyester	131,967b	2512b	0	0

[z]Mean separation within columns by Duncan's multiple range test, 5% level
Source: Wells and Loy 1985

are industrial polymers which have been adapted for agricultural use, rather than polymers that were designed for specific row cover application.

It is important to remember that the purpose of row covers is to increase productivity through an economical increase of early and total production per unit area of growing area. However, row covers alone will not meet this objective.

They are only one segment of the overall production system. Maximum productivity and financial return will be realized only through the efficiency of grower management and marketing.

4
GREENHOUSES

A greenhouse has basically one purpose, that is to provide and maintain a growing environment that will result in optimum crop production at maximum yield.

BASIC CHARACTERISTICS

Enclosure. As a structure for growing plants, greenhouse must admit the visible light portion of solar radiation for plant photosynthesis and, therefore, must be transparent. At the same time, to protect the plants, a greenhouse must be ventilated or cooled during the day because of the heat load from the radiation. The structure must also be heated or insulated during cold nights. A greenhouse acts as a barrier between the plant production areas and the external or general environment. Production is protected from external stresses such as weather (wind, hail, drought, rain, frost, etc.) and pollution from industrial and other sources (acid rain, particulates, etc.).

Internal Control. Greenhouse agriculture is often termed controlled environment agriculture (CEA) which implies control over the internal environment. Through careful control of light, carbon dioxide (CO2), humidity, temperature, water content of growth medium and nutrient levels, a CEA facility may be managed for optimum production levels. These factors can be altered for different crops to regulate production schedules and in response to pest and disease attacks.

Product Yield. A greenhouse, or CEA, may produce yields that are dramatically higher than those of open-field agriculture (Table 12) which may result higher per hectare gross revenues.

Table 12. Comparative yield for a single crop grown in Abu Dhabi greenhouses vs. field grown

Type of Vegetable	Abu Dhabi greenhouse MT/ha	Good yield field grown MT/ha	Number of crops per year in a greenhouse
Broccoli	3.25	10.5	3
Cabbage	11.5	7.5	4
Cucumber	57.5	30.0	3
Eggplant	28.0	21.0	2
Pepper	32.5	15.6	2
Tomato	150.0	75.0	3

Source: Jensen, 1977.

The yield increases result from:
- An increased number of crops per year. For example, the number of crops that could be grown in the Abu Dhabi greenhouses (Table 12) far exceeded open field opportunities. In most cases the total greenhouse yield per year was doubled, tripled, or even quadrupled as compared to open field yields. Those grown outdoors were subject to the high temperature extremes of summer or the winter cold, which is especially dangerous to heat-loving crops such as cucumber, eggplant, and pepper.
- An enhanced growing environment (faster growth, longer harvest) allowing plants to come closer to their photosynthetic capabilities.
- Diminished losses to weather and other stresses.
- Plant breeding for yield, with an environment tailored to plant requirements.

Reliability. Because of increased control over the factors of production, CEA production can be highly reliable. Production quantity, quality, and schedule are quite predictable; the wide and erratic fluctuations of open field production can be reduced significantly . CEA operation is, however, dependent on an assured supply of material inputs: it is wholly dependent, for example, on stored water or water supplied from external sources.

Scheduled and Continuous Production. In CEA, the schedule of production can be made relatively less dependent on season than open field agriculture. The production schedule can be planned to take advantage of general market needs. A series of greenhouse modules can be coordinated for continuous operation and production.

Quality of Produce. Assuming adequate knowledge of plant science and plant nutrition, CEA can allow the growth of floricultural crops and vegetables that are of consistent quality. Unblemished vegetable crops of reliable taste, size, and texture will represent a further advance in quality. CEA presents an opportunity to ship ripe produce to adjacent markets, instead of green produce to remote markets.

Input Conservation. CEA requires lower recurring input levels per unit of produce than open field agriculture. This is due principally to:
- the lower dissipation rates that result from restricting the flow of the dissipation media (runoff water, air, and soil erosion);
- the capability of end-of-pipe recovery and recycling of water and fertilizer;
- the spatial concentration of production (with higher ratios of exposed surfaces to input diffusion space); and
- increased control over application.

These savings can be offset, however, by higher initial material requirements.

Regional Location Flexibility. Physically, if not always economically, the regional location of CEA is considerably more flexible than for open field agriculture. Climate and weather conditions exert less control over production, and less usable land is required per unit of produce. In fact, with proper CEA system design and management, food may be grown in CEA anywhere in the world, even in outer space.

Urban-Rural Location Flexibility. Because of CEA's much higher yields (which greatly reduce the amount of land required), agricultural production may be either in rural or near-urban areas.

Non-Food Systems Integration. CEA can be integrated with other systems, such as energy utilities, sewage disposal, and water supplies. This is partly due to CEA's point-source input and output characteristics, its location flexibility, and the physical concentration of the facility. The wastes of one system may provide the inputs for the other; and shared facilities may lower the initial operating costs of both.

The technological status of CEA today includes a number of innovative schemes covering imaginative efforts to locate CEA in arid and tropical areas as well as temperate regions, to use desalted sea water, to produce fish as well as plants, to automate CEA operations, and to use both liquid and solid growth media.

SITE SELECTION

There are many factors for consideration in determining the amount of greenhouse space to build. They are: investment capital available, management skills and training, type of business - wholesale/retail, crop selection and their environmental requirements, market, labor requirements, and personal preferences.

A good building site is crucial to the function and operation of a greenhouse. A site should not be in the shade of trees, mountains, or even clouds. In tropical regions, afternoon clouds often bank against a mountain, which causes severe light obstruction to crops in a greenhouse. In northern latitudes, a slope facing south is good for winter light and protection from northerly winds. Sites with chronic winds should be avoided, especially in cold regions: winds substantially increase the heating requirements, and wind breaks can only be installed if they do not obstruct light. In deserts, where blowing sand is common, sand often collects in the region of the greenhouses; therefore, such sites should be avoided. Good access is important in site selection in regions with heavy snowfall.

The selected location should provide for adequate soil drainage. Considerations include ground slope, the drainage of surface water, and subsurface drainage; the latter may require digging test holes to investigate existing or potential problems.

The availability of a dependable and economically efficient source of energy is also vital to site selection. Greenhouses may require electricity and fuel for heating and mechanical cooling. An electric power distribution line adjacent to the site reduces the investment needed to bring the electricity to the greenhouse. If a dependable source of electricity is uncertain, a stand-by electrical generator is essential, especially if electricity is needed to control the greenhouse environment, for irrigation, and for storage of horticultural products. A short access road to an all-weather road assures fewer problems in maintaining adequate fuel and greenhouse supplies. The access road is also critical for the transport of greenhouse products to the market, especially one that does not cause damage to the product while in transit. Telephone service is also necessary for a successful operation.

A dependable supply of good quality water is absolutely necessary for a successful greenhouse operation. A groundwater geologist and/or a local well driller should be consulted to determine the potential for an adequate water source. A water sample should be analyzed for its agricultural suitability. The water must be free of heavy metals, low in extraneous salts, such as sodium, and low in boron or any other elements that might cause phytotoxicity to plants if allowed to accumulate in the growing medium.

In some countries, zoning regulations may control the use of land; therefore one must consult the appropriate government agencies before planning the facility. Site selection should consider the possibility of facility expansion.

STRUCTURAL DESIGN

There are many types of greenhouse structures used successfully in protected agriculture. Although there are advantages of each for particular applications, in general, there is no one best greenhouse.

The structural design of a greenhouse must provide protection against damage from wind, rain, heat, and cold. At the same time, the structural members of a greenhouse must be of minimum size in order to permit maximum light transmission to the crop.

Design loads for a greenhouse structure include the weight of the structure itself and, if supported by the structure, the heating and ventilation equipment and water lines. The load may also include the weight of crops trained to a support system carried by the greenhouse frame, and loads from wind and snow. Greenhouse structures should be designed to resist a 130 km/hr. wind. The actual load depends on wind angle, greenhouse shape and size, and the presence or absence of openings and wind breaks.

Frame Materials. Wood, bamboo, steel, galvanized steel pipe, aluminum, and reinforced concrete are all materials used to build frames for greenhouses. Frames often incorporate a combination of materials. Wood must be painted white to improve light conditions within the greenhouse, but care should be taken to select a paint that will inhibit the growth of mold. Wood must also be treated for protection against decay. It is especially important to treat, with preservatives, any wood that may come into contact with the soil. Treatment must be free of chemicals that are toxic to plants or humans: this eliminates woods treated with creosote and pentachlorophenol, which must not be used. Chromated copper arsenate (CCA) and ammonical copper arsenate (ACA) are water-borne preservatives that are safe to use where plants are grown. Even wood materials, such as redwood or cypress, with natural decay resistance should be treated, especially in desert or tropical regions. Unfortunately, these woods are becoming more difficult to obtain at prices competitive with other materials.
While use of reinforced concrete is generally limited to foundations and low walls, concrete is sometimes used in the People's Republic of China as support posts for a frame made of bamboo. Most often, frames may be all aluminum or steel or a combination of the two materials. Aluminum is comparative-

ly maintenance-free as is hot dipped galvanized steel. In tropical areas, it is advisable to double dip the steel, especially if the single dip galvanizing process does not give a complete cover of even thickness to the steel. Aluminum and steel must be protected from direct contact with the ground to prevent corrosion. If there is a danger of any part of the aluminum or steel coming into contact with the ground it must be thoroughly painted with bitumen tar.

Structural Form. A straight sidewall and an arched roof is possibly the most common shape for a greenhouse; the gable roof is also widely used (Figs. 8 and 9). Both structures can be free

standing or gutter connected (Figure 10) with the arch roof greenhouse.

The arch roof and hoop style (Figure 11) greenhouse are most often constructed of galvanized steel pipe bent into form by a roller pipe bender. In tropical areas, bamboo is often used to form the gable roof of a greenhouse structure.

If tall growing crops are to be grown in a greenhouse or benches are used, it is best to use a straight side wall structure rather than a hoop style house; this ensures the best operational use of the greenhouse. A hoop greenhouse is suitable for low-growing crops such as lettuce, or for nursery stock that might be housed through the winter in a greenhouse located in an

Figure 8. A straight sidewall greenhouse structure with arch roof

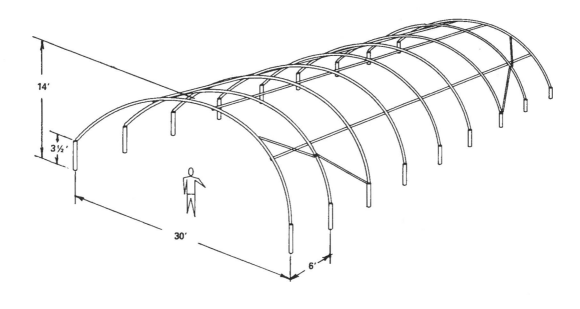

Figure 9. A straight sidewall greenhouse structure with a gable roof

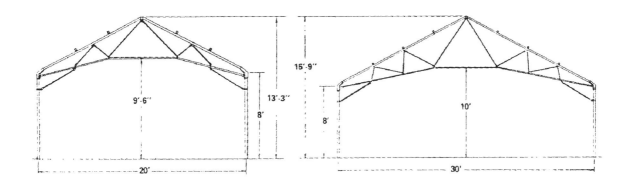

Figure 10. A gutter-connected straight sidewall greenhouse

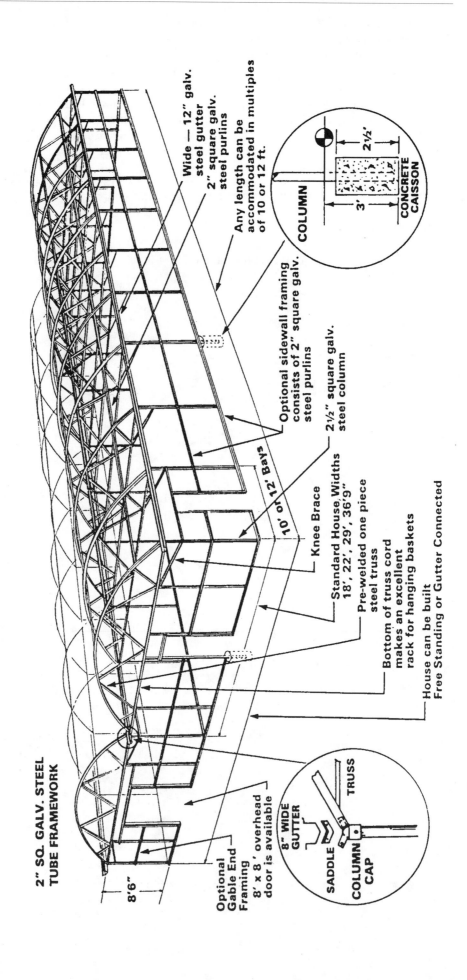

Wide — 12" galv. steel gutter

2" square galv. steel purlins

Any length can be accommodated in multiples of 10 or 12 ft.

COLUMN

2½'

3'

CONCRETE CAISSON

Optional sidewall framing consists of 2" square galv. steel purlins

2½" square galv. steel column

10' or 12' Bays

Knee Brace

Standard House Widths 18', 22', 29', 36'9"

Pre-welded one piece steel truss

Bottom of truss cord makes an excellent rack for hanging baskets

House can be built Free Standing or Gutter Connected

2" SQ. GALV. STEEL TUBE FRAMEWORK

8'6"

Optional Gable End Framing 8' x 8' overhead door is available

8" WIDE GUTTER

TRUSS

SADDLE

COLUMN CAP

Figure 11. A hoop style greenhouse

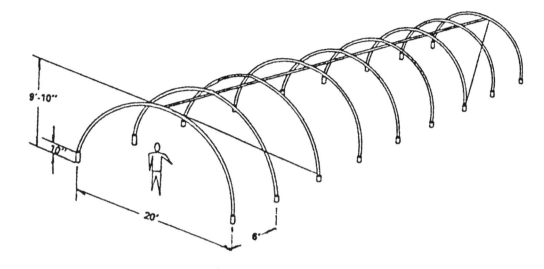

Figure 12. A gothic arch frame greenhouse

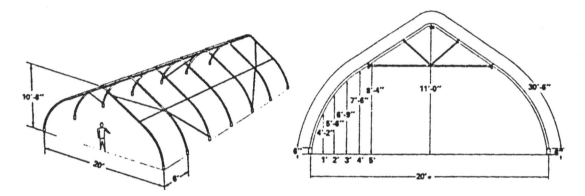

extremely cold region. A gothic arch frame structure (Figure 12) can be designed to provide adequate sidewall height without loss of strength to the structure. This form of structure, as with others, can be used as a single free-standing greenhouse or as a large range of multi-span, gutter-connected units.

Air supported greenhouses are used at times as temporary structures but are not recommended for permanent installation because of problems of entry, warm weather ventilation, and possible power loss.

Greenhouse Covering. Glass is still a common glazing material. Large panes reduce the shading of crops from the glazing bars. Dutch greenhouses have panes extending from the valley to the ridge. These houses have the advantages of few parts and ease of construction. Large panes, bar caps and strip caulking material have reduced the labor required for glass glazing.

Despite the dominant position of glass as a covering for protected structures in Northwest Europe, glass remains inflexible, heavy, and expensive. Consequently, the hectarage of glasshouses on a world basis has remained static (approximately 30,000 ha.) during. the last 25 years. In contrast, the quantity of plastic used for greenhouses is increasing rapidly.Since polyethylene sheet film was first developed in England in 1938,

it has been used widely in greenhouses because it is easy to work with and inexpensive. These structures are primarily used to increase temperatures during the winter and to protect against wind. In Northwest Europe plastic greenhouses are replacing cold frames, glass cloches and single span glasshouses, but not gutter connected glasshouses.

Plastics, other than polyethylene, have been used for greenhouse glazings. Polyvinyl chloride (PVC) film has a very high emissivity for long wave radiation (similar to glass), which creates slightly higher air temperatures in the greenhouse during the night. The Japanese consider this improvement in thermal environment a benefit that outweighs the price advantage offered by the less expensive polyethylene (PE). The disadvantage of PVC is its narrow width as compared to PE, which may be manufactured to a width of approximately 15 meters. The narrow PVC sheets can be heat-welded together to form a large sheet, but this adds to the cost of the glazing material.

The large sheets of PE can be applied as an air-inflated "blanket" over a greenhouse: two sheets of PE are separated by air pressure maintained by a small continuously running fan. This arrangement provides approximately 30-40% heat savings during winter. The double-layer, air-inflated roof has also proven valuable in regions with high winds or typhoons. It offers stability during these periods, saving the greenhouse and the crop during times when structures covered only with one layer of plastic are often lost and the crops damaged or destroyed. PVC film is not suitable for air-inflated roofs because the air pressure stretches the film and damages its structural strength. Because PVC film is not photodegradable, as is PE, environmental concerns about disposal may diminish the use of PVC in Japan in favor of PE.

Other plastics have been used for greenhouse glazings, but with indifferent technical or economic results. Corrugated fiberglass panels, polyester (Mylar), polyvinyl fluoride film (Tedlar), acrylic polyester film, etc., have proved either more unsuitable, inconvenient, or, in most cases, expensive than polyethylene, even though the latter may have to be replaced more frequently. New materials — polycarbonates, acrylics, double-skinned panels, light-selective films laminated to polyethylene or Tedlar, etc. — are either recently in use or currently being studied. At present, however, their technical merit is offset by the high costs, and makes them unsuitable for use in developing agricultural communities.

Greenhouse designs in Holland use large panes of glass, allowing for rapid glazing of the greenhouse structure.

The ideal greenhouse "selective film" should do the following (Anon. 1977):

1. *Transmit. the visible light portion of the solar radiation spectrum, the only portion utilized by plants for photosynthesis.*
2. *Absorb the small amount of ultraviolet in the spectrum (3~) and cause some of it to fluoresce into visible light, useful to plants.*
3. *Reflect or absorb infrared radiation (~9~ of the spectrum), which plants cannot use and which causes greenhouse interiors to overheat.*
4. *Minimize cost, and have a 10- to 20-year useable life.*

Such a film would obviously improve CEA performance, increasing light levels and crop yields and reducing solar heat load. The first three criteria present no extraordinary technological challenge. However, in the context of the small American greenhouse industry, the inventiveness of plastics manufacturers has been inhibited by their understandable doubts about the market potential for such a product.

ENVIRONMENTAL CONTROL

Light. One of the major determinates of plant growth is light, both outdoors and in protected agriculture. The three light-requiring processes that govern plant growth are:

a. photosynthesis
b. photomorphogenesis
c. photoperiodism

Of these, photosynthesis, in which radiant energy is converted into the chemical energy necessary for the synthesis of carbohydrates, is the most important. Photomorphogenesis is the formative effect of light, which regulates germination, inhibits

excessive stem elongation of dark-grown seedlings, promotes the enlargement of cotyledons and leaves, and starts the production of chlorophyll, turning the color of the leaves from pale yellow to green. Photoperiodism includes several phenomena that occur in plants which depend on the length of the daily light period; such phenomena involve flowering, winter dormancy, and generative reproduction.

Since light is essential to plant growth, it is important in greenhouse construc~ion to keep structural material to a minimum, so that incoming light will not be blocked. This is especially important in designing for vegetable crops grown in northern latitudes, which require maximum amounts of incoming light.

Photoperiod (the length of the light and dark period) strongly affects the flowering process of many greenhouse plants. For example, chrysanthemums are short day plants, requiring 9-12 hours of light to flower. Therefore, greenhouses are often lighted during the winter to induce flower formation. In northern Europe, it is economically viable, in many instances, to grow vegetable transplants or ornamental plant seedlings under artificial light in the greenhouse nursery durins winter. In these countries, it is also common to provide additional light to bring flower crops to full maturity.

Light can come in many forms: incandescent, tungstenhalogen, fluorescent, and high intensity discharge lamps (HID).

High output HID lamps are compact and are increasingly popular for greenhouse and growth chamber lighting. They are efficient producers of photosynthetically active radiation (PAR) light. HID lamps may be mercury, metal halide, high pressure sodium, or low pressure sodium.

Uniform distribution of light is impossible without reflectors. Reflectors should be designed to direct the light uniformly over the plant area.

The cost of lighting equipment, and its operation make artificial lighting extremely expensive, and is, therefore, a serious consideration in determining the economic advisability of pro-

Figure 13. Continuous air recirculation and motion.

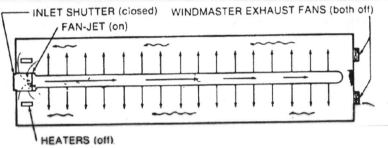

INLET SHUTTER (closed)   WINDMASTER EXHAUST FANS (both off)
FAN-JET (on)
HEATERS (off)

within the crop zone. Natural convection air movement will develop in a closed greenhouse when warm, light air rises, is cooled against the roof surfaces, and then falls back to the ground to be warmed. The result is a warm air mass in the attic of the greenhouse and large temperature variations occurring between the upper part of the greenhouse and ground level. Uniform temperatures can be obtained with the use of small fans to create a slow horizontal movement of the air mass. In greenhouses containing plants that have tall, dense foliage such as roses, tomatoes, or cucumbers, air movement can be provided by using perforated polyethylene film ducts to discharge air within the crop (Figure 13). To assure relatively uniform distribution, total fan capacity should be equivalent to moving about one-quarter of the greenhouse volume per minute.

Ventilation/Cooling Systems. Ventilation requirements for greenhouses vary greatly, depending on the crop grown and the season of production. The ventilation system can be either a passive system (natural ventilation) or an active system using fans for forced ventilation. Greenhouses used seasonally usually employ only natural ventilation; those in year-around production, especially in arid regions of the world, use fan ventilation. Evaporative cooling is often part of a fan ventilation system.

Ventilation/cooling equipment includes such items as vents, fans, shading, and evaporative pad systems, as well as control components. Passive ventilation on glass structures consist of ridge and side vents that are opened or closed manually or by modernized vent thermostats. With plastic greenhouses, passive or natural ventilation is a key challenge, especially without the aid of exhaust fans.

One of the simplest methods of ventilation is to roll up the sides, allowing air to flow across the plants. Narrower greenhouse tunnels, (i.e. 4.5 meters) are easier to manage than 7 meter wide tunnels. Roll-up sides are easily accommodated by attaching a permanent board or strip (such as 0.4 cm by 2 cm) on the bows (spaced 1.3 m apart) about 1 meter up from the ground (Wells 1991). By using a batten strip, a plastic cover can be permanently attached on the sides below the permanent board. The upper part of the sheet is attached to the board and the bottom is buried in the soil. A 0.4 cm pipe is then taped to the lower edges of the plastic sheet covering the tunnel. The pipe, with a simple hand crank attached to one end, facilitates rolling up of the sides to any desired height. The amount of ventilation on one side, or both sides, may be easily adjusted in response to temperature, prevailing wind, and rain. During periods of excessive heat, it may be necessary to roll the sides up almost to the top of the tunnel. When the structure is not in use, the plastic may be removed to prevent unnecessary degradation by ultraviolet light. If this is done, a UV inhibited plastic cover may last for a period of up to 4-5 years.

Passive ventilation can also be accomplished by manually lowering the side wall of the greenhouse with a curtain winch similar to that used on a boat trailer, or by simply raising or parting the polyethylene sheeting manually. On those tunnels

ducing a particular crop. In arid or tropical regions, artificial lighting systems are rarely used. In northern latitudes products grown under expensive lighting systems rarely face any competition from imports, since transportation and handling charges also inflate the cost of these products. These systems are economical only where consumers' buying power is strong enough to absorb the high cost of imports or of products grown under controlled environment/artificially lighted agricultural systems. Unless there is consumers' buying power, out-of-season products will not be available, as is the case in many developing countries.

Temperature Control.

Heating. While a greenhouse can be expensive to heat, it can be profitable if high quality, cost competitive products are grown.

The selection of heating equipment depends on the size and type of operation, the greenhouse structure, fuel availability and cost, and the cost of the system componentsi The fuel can be gas, oil, coal, or wood. A heating system consists of a fuel burner, heat exchanger, distribution system and controls. The fuel burner can be located within the greenhouse with the heat delivered to the crops by convection and radiation. These direct-fired units use either air or water as the heat transfer fluid.

Central boilers are often used in large operations where gutter-connected greenhouses are used. Either water or steam acts as the heat transfer unit. Most steam systems use a low pressure boiler, with enough pressure to push the steam to all greenhouses in a range.

Air movement in a greenhouse is needed for acceptable carbon dioxide distribution and to maintain uniform temperature where ventilation is provided by raising the plastic sheeting, the plastic is lifted at ground level, at each bow, and held in position by friction created by cords or small lines placed over the plastic cover between each structural bow. Each cord has a stretch line, or piece of rubber, to provide flexibility in the line which accommodates lifting of the plastic for ventilation but provides enough tension to prevent damage by wind.

Opening the ends of a greenhouse tunnel does not provide sufficient ventilation, no matter what the length of the tunnel. Ventilation must be provided by side vents as well, especially during warm periods of the growing season.

If insects are common, especially those which are vectors for virus diseases, the open vent areas must be covered with screens.

Such ventilation systems on plastic greenhouses are only effective on free-standing greenhouses and not on gutter-connected structures. An exception is a sawtooth gutter connected greenhouse illustrated in Figure 14; an eight meter wide shelter type structure designed to utilize natural airflow for cooling.

A new concept in natural ventilation for quonset (see Glossary) greenhouses incorporates a one-meter continuous vent into the roof along the entire length of a gutter-connected greenhouse. The vent can be operated with a modernized vent thermostat for automatic climate control or by a controller/computer system.

In the tropics, the sides of greenhouse structures are often left open for natural ventilation. A tropical greenhouse is primarily a rain shelter, a cover of polyethylene over a crop to prevent rainfall from entering the growing area and, in turn, mitigate the problems of foliage diseases. To prevent insects from entering, especially those which are vectors for virus diseases, the side areas are covered with screens. The use of these non-chemical means of insect control become increasingly important as concerns mount about the long-term effects of chemicals entering the food chain or the environment, and the exposure of workers to toxic compounds. Screens must have holes

Roll-up sides on a polyethylene greenhouse is a simple method to provide natural ventilation during periods of warm weather.

large enough to permit free flow of air; screens with small holes block air movement and foster a build-up of dust. Greenhouses can be cooled by using the natural forces of wind and temperature; however, a mechanical system is required for fine temperature control.

For active or mechanical ventilation, low pressure, medium volume propeller blade fans, both directly connected and belt driven, are used for greenhouse ventilation. They are placed on the end of the greenhouse opposite the air intake, which is normally covered by gravity or motorized louvers. The fan vents, or louvers, should be motorized, with their action controlled by fan operation. Motorized louvers prevent the wind from opening the louvers , especially when heat is being supplied to the greenhouse. Wall vents should be placed continuously across the end of the greenhouse to avoid hot areas in the crop zone.

Evaporative cooling in combination with fans is called *fan and pad cooling*. The cooling pads can be made from a number of materials; most often they are made of a cellulose material, usually aspen wood, or a multi-celled/honeycombed material called "kool-cel." Evaporative cooling systems are especially efficient in low humidity environments. There is increasing interest in building greenhouses combining both passive and

Figure 14. A sawtooth greenhouse for natural ventilation

"Sawtooth Greenhouse"

NATURAL AIR FLOW   NATURAL AIR FLOW   NATURAL AIR FLOW

Source: Stuppy Greenhouse Mfg., Inc.

Mechanical ventilation provided by propeller blade fans are commonly used in combination with evaporative cooling.

Although high fuel costs may be counterbalanced, to a degree, with new crops or cropping techniques and higher prices to the grower, they will not necessarily offset the continuing rise in the cost of fuel. Since 1972, the cost of greenhouse heating with natural gas in the United States has risen three to four times. Many growers using natural gas are under a curtailment or allocation policy, and are often forced to supplement their fuel needs with more expensive fuels. If they have heating systems designed to use only natural gas, they may need to reduce their growing area or length of growing season. The cost to heat with oil in the State of Ohio may reach over $100,000 per hectare, a cost that has forced many greenhouse vegetables being raised in the Cleveland, Ohio area and its surrounding counties; now that industry is below 60 hectares.

Because of the lesson of the 1970's oil embargo, and the rapid increase in energy cost, the research community throughout the world, along with greenhouse growers, have developed many new concepts and ideas in energy conservation and alternatives. Today, many of the following systems are commonplace in the greenhouse industry.

active systems of ventilation. Passive (natural) ventilation is utilized as the first stage of cooling: fan-pad evaporative cooling takes over when the passive system is not providing the needed cooling. At this point, the vents for natural ventilation are closed. Initial costs of installation are greater when both options for cooling are designed into greenhouse construction. Even so, long-range operational costs are minimized, since natural ventilation will, most often, meet the needed ventilation requirements.

Fogging systems can be an alternative to evaporative pad cooling. They depend on absolutely clean water, free of any soluble salts, in order to prevent plugging of the mist nozzles. Such cooling systems are not as common as evaporative cooling pads but, as they become more cost competitive, may be become used more frequently. Fogging systems can be the second stage of cooling when passive systems are inadequate.

Energy Conservation and Alternatives. Investment in energy from petroleum and electricity permits the production of food and floral crops out of season at yields and quality often far superior to those grown outdoors. However, control of the environment within a greenhouse may require large amounts of energy, and energy is, therefore, a prime factor in computing profitability. Furthermore, some fuels raise the spectre of dependence that may threaten greenhouse production. In the early 1970's, the United States bought half of its oil from abroad. Dependence on foreign oil has increased dramatically since that time, amplifying the threat to today's greenhouse industry in the United States, particularly non-food crops, should another oil embargo occur.

Thermal Conservation. Significant energy savings can be made in a greenhouse if a grower implements the following suggestions by the American Council for Agricultural Science and Technology (CAST 1975): 1) tighten up the house, closing all possible openings, 2) use polyethylene or fiberglass on the inside of gable ends, 3) maintain the steam or hot water system regularly to stop leaks, 4) use reflector materials behind heating pipes to reflect heat out into the greenhouse, 5) maintain the boiler for operation at peak efficiency, 6) insulate hot water and

Propeller ventilation fans are often placed in the roof of the structure in the middle of the greenhouse. Outside air is drawn through cooling pads on both ends of greenhouse.

Figure 15. A typical greenhouse layout covered with multi-layers of plastic

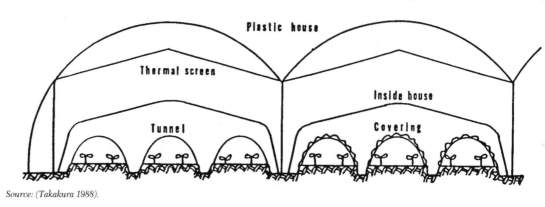

Source: (Takakura 1988).

steam supply and return piping, and inspect at intervals, replacing the insulation when needed, 7) maintain the automatic valves in the heating system, and 8) check the thermostats regularly for proper operation.

Covering a greenhouse with a double layer of polyethylene to reduce the loss of heat energy was first reported over three decades ago by Sheldrake and Langhans (1961). Roberts and Mears (1969) were the first to work out the construction details and inflation requirements for the double-layer, air-inflated roof concept, a technique now used worldwide. Fuel savings up to 40% (Axlund et al., 1974) have been observed in New York and California.

In Japan growers place a removable sheet of polyethylene over the crop and a row cover over each plant row in order to reduce heat loss from the greenhouse during the night (Figure 15). The plastic row cover and inner polyethylene covers are pulled to the side during the day to maximize incoming light. The inside row covers are used primarily for growing seedlings and are normally removed when the crop grows to maturity. Opaque sheets also can be applied at night. Floating mulches are becoming increasingly more popular as an alternative to inside row covers. According to Takakura (1988), more than 90% of the heated greenhouses have at least one layer of movable thermal screen.

In Germany (Strickler 1975), a system made of polyethylene tubing is installed which, when the tubes are inflated, seals the growing area from the roof surface area. Insulation is effected by two thicknesses of polyethylene. Including the volume of still air, the system reduces heat loss by 35 percent. The tubes may be inflated at night and deflated during the day, except during periods of bright sunlight: at these times they may be left inflated to provide shade for crops such as foliage plants.

At Pennsylvania State University (White, et al. 1976); twenty different materials were laboratory tested for their potential as heat loss barriers during dark periods. Foylon, a porous 44 x 55 count (5 oz.) polyester, aluminum foil hybrid fabric, reduced the amount of fuel used in a glasshouse by 57%. Such internal fabric curtains are certainly effective in reducing heat loss. Improvements in curtain materials, resulting in greater reflective properties, should increase their energy conservation potential while permitting their use for photo period control or shading purposes.

Interior curtains have two drawbacks; especially in small greenhouses: storage of the curtains when open, and potential for augmenting snow accumulation. Unless they are out of the way when not needed, curtains will shade plants and make personnel movement inconvenient around the interior perimeter of the greenhouse. The initial installation of curtains in an existing greenhouse could be awkard, depending on the structure of the greenhouse and the configuration of the curtain volume. Heavier snow loads might also develop with the use of interior curtains. Since temperature of the glass will generally be lower than without the curtain, snow may accumulate for a longer period of time, even beyond the design load of the greenhouse. This would endanger the structure, especially if high wind loads also occur (White et al., 1976). Despite these problems and the relatively high initial costs, over a long period of time, internal curtains can be one of the most practical and economical methods of energy conservation.

Most energy conservation methods can be classified as either modification or maintenance techniques. A few methods, also important, fit into neither category. Modification and maintenance methods apply to both the structure and the heating system. Structural modifications usually reduce infiltration and add insulation to the

In Japan, crops such as cucumbers are commonly covered with several layers of plastic to provide additional protection against cold nights. During the day, the row covers and inner greenhouse cover are pulled to the side.

Movable curtains provide a practical and economical way to conserve energy during nighttime heating.

greenhouse. Heating system modifications seek to optimize the recovery of heat from burned fuel.

Table 13. Summary of potential annual savings for energy conservation methods (Badger and Poole 1979)

Method	Annual % Saving Range
Glass	0 (base)
A. Major Modifications	
Continuous	
1. Double plastic film over glass	40-60
2. Glass lap sealants	5-40
3. Single plastic film over glass	5-40
4. Double layer plastic film	30-40
Periodic	
5. Curtains	20-60
6. Polystyrene pellets	60-90
7. Liquid foam	40-75
B. Other Modifications	
1. Sidewall insulation	5-10
2. Foundation insulation	3-6
3. Insulating ventilation fans	1-5
4. Heating systems	
a. Automatic firetube cleaners	6-20
b. Turbulators	8-16
c. Stack heat recovery unit	?
C. Maintenance	
1. Structure	3-10
2. Heating system	10-20
D. Miscellaneous Factors	
1. Windbreaks	5-10
2. Greenhouse orientation	5-10

The energy conservation methods in Table 13 are listed for ease in gathering potential cost and fuel savings. Table 13 is based on the standard glass greenhouse as a reference.

"Continuous" methods indicate 24 hour use; "periodic" methods refer to nighttime insulation only. The actual annual savings realized will vary with location, type, and conditions of the greenhouse, and weather conditions. It is possible, with ideal conditions, to observe savings greater than those listed. The figures must be carefully evaluated for unknown conditions, especially if based on short time intervals (Badger and Poole 1979).

Solar Energy. The potential of solar energy for greenhouse heating received great attention in the early 1970's when the cost of oil started to increase rapidly. Greenhouses are inherently good solar collectors when sunshine is available; in fact, excessive daytime heat is often a problem. On the other hand, greenhouses can have a high thermal loss at night, when over 75% of all supplemental greenhouse heating is required. Therefore, solar heating may not be a total substitute for greenhouse energy needs. Indications are that it would not be economically feasible to provide for 100% substitution of the greenhouse energy requirements with solar energy; however, up to an estimated 80 percent of these needs may be satisfied by solar energy. Most of the application problems are distinct from those encountered in heating and cooling of resioences or office buildings.

Studies were conducted at Cornell University using the greenhouse itself as a solar collector rather than through use of external collectors (Price et al. 1976); this saves the cost of additional collectors and the equipment necessary to transfer the heat to the greenhouse. Unfortunately this application is relatively inefficient for greenhouse production since plants cannot stand the high temperatures necessary to store large amounts of heat energy. In addition, at present it is not practical to tilt a greenhouse to the south at steep angles to collect the energy

Figure 16. Concentrating hot water solar collectors

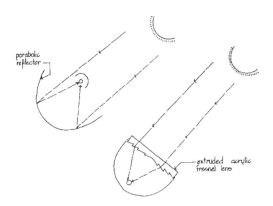

Source: (Peck, 1976)

efficiently. For these reasons, many investigators have directed their efforts to greenhouses heated by collectors separate from the greenhouse.

Basically there are two major classifications of external solar collectors: concentrating collectors and flat-plate collectors (Peck 1976). Concentrating collectors can involve reflective mirrors that concentrate the sunshine on a line or point (the reverse of a car headlight). The intensified sunshine strikes a target, which might be a pipe containing a moving fluid, such as water (Figure 16).

Very high temperatures can be attained in this way. However, most of these devices must be moved to a certain extent to track the sun. Also, because they utilize only the direct sunlight, their performance is quite low on hazy days. Direct sunlight forms a shadow when it strikes an object, and its apparent source is the sun; while diffused sunlight has its source in the hemisphere of blue sky and does not form a shadow.

Another type of concentrating collector uses an extruded acrylic plastic Fresnel lens that concentrates sunshine on a pipe below. This collector is simply an ordinary convex lens compressed onto a flat sheet; it resembles the headlight on a car. Because concentrating collectors do not appear to be economically feasible for greenhouse heating, they are not being considered for use to any great extent.

Flat-plate collectors, as the name implies, are flat sheets of glass or plastic through which sunshine passes. The sunshine is absorbed by the dark surface beneath the glass and is converted to heat. Heat is lost mainly through the clear materials that cover the dark surface. These devices can also be tilted to face the sun squarely during the en-tire day, but this is a complexity that is generally avoided. The tilt, with respect to the horizontal plane, controls the amount of heat collected. Hence, in placement it is important to ascertain the most effective tilt of the collector relative to the sun. Most collectors face due south, although they may vary by as much as 10_ to 20_ to the east or west.

There are two types of flat-plate collectors those in which water or a water mixture is the collection fluid (Figure 17) and those in which air is the collection fluid (Figure 18).

At the Lockheed-Huntsville Research and Engineering Center, In Alabama, hot water flat-plate solar collectors were tested to heat a greenhouse (McCormick 1976). A collector shed housed the instrumentation, storage, pumps, and controls. This system was designed to provide 75 percent of the energy needed from the solar collectors, but actual performance exceeded expectations, providing approximately 80 percent of the heating needs of the greenhouse. The remaining 20 percent of the energy needs were provided by an auxiliary system using fossil fuel.

The Lockheed tests indicate that significant amounts of heat for night in a greenhouse can be supplied by solar energy, but they point out that the performance and technical simplicity of solar heating systems for greenhouses disguise a very difficult economic problem. Because of the relatively short utilization period for a greenhouse solar heating system, only a fraction of the year-round available solar energy may be collected and used. This causes a much reduced yearly return on the initial investment compared to home heating and water heating, for example. In most parts of the United States, a solar collector is

Figure 17. Flat-plate hot water collector

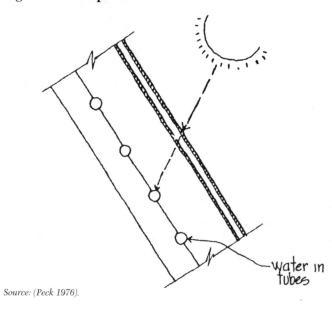

Source: (Peck 1976).

Figure 18. Fin type hot air collector

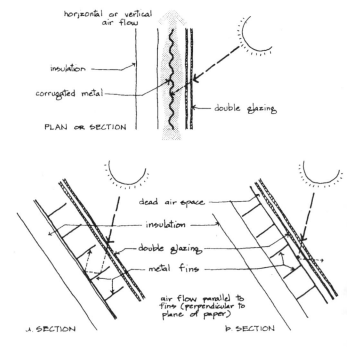

Source: (Peck 1976).

worth only about $4.30/m^2 per year for heating greenhouses. This means that a system cost of more than about $53.80/m^2 of collector area will require a prohibitively long time to recover the initial investment. However, typical expenses are more than this for the collector alone, not including heat storage, pumps, etc. This seems to rule out prefabricated collectors under present conditions.

Research at Rutgers University has involved the design of collectors of low initial cost. Such collectors are made of polyethylene and are seemingly well suited for use with polyethylene film greenhouses (Roberts et al. 1976). Figure 19 illustrates the basic

Figure 19. Cross section of a sloped flat-plate hot water solar collector made of polyethylene

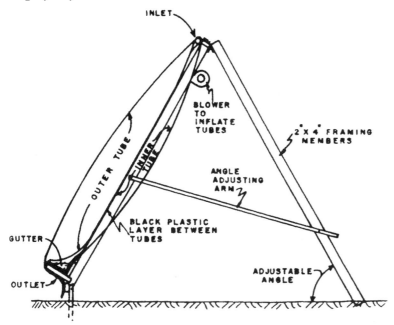

Source: (Roberts et al. 1976).

design of the warm water solar collector which is 3 m high and 7.3 m long. It is designed so the angle can be adjusted. Four layers of 150 micron clean ultraviolet stabilized plastic and one black layer are arranged to provide structural stability and some insulation. The black layer is sandwiched between two air-inflated sections of clear film. Water is pumped to the top of the collector where it leaves the 3 cm pipe through holes spaced 10 cm apart. Water flows down between the inner layer of the front inflated section and the black layer, is collected in a trough and returned to storage. Sheet flow occurs by forcing the clear film against the black film by a static pressure of about 0.64 cm of water. A small quantity of detergent added to the water increases sheet flow. Flow rates have varied from 1.9 to 2.5 liters per minute per 30.5 cm of collector. Temperature records obtained on a clear November day are shown in Table 14.

From these data, it is evident that, for a modestly insulated collector, efficiency decreases with increasing water temperature due to an increase in heat loss. The total insulation available for this five hour period on November 28, 1975, was 3,252 K Cal/m^2. Assuming an average collector efficiency of 55%, the energy absorbed by the collector was 39,798 K Cal. Outside ambient temperatures ranged from 15 to 20°C. The collector has withstood winds of over 96 kph and several snow storms without damage. The lumber, structural supports, hardware, and plastic film costs for the 3 x 23 m collector were $120 or

$1.61/m^2 of collector. The annual cost of film replacement would be $1.61/m^2.

Among the many disadvantages of external solar collectors is the difficulty of integration into a greenhouse location, especially in large greenhouse complexes (Price et al. 1976.) An external collector located too close to a greenhouse will shade the greenhouse or the greenhouse may shade the collector. A collector too distant from the greenhouse may result in excessive, heat loss in transit, friction pressure drop and material expense. Distant placement also requires increased allocation of land space since a 1,000 m^2 greenhouse requires at least 500 m^2 of collector. Another disadvantage is that covering a greenhouse with heat collecting (opaque) panels eliminates the light essential for proper plant growth. Prefabricated collectors made of long-lasting materials are prohibitively expensive; polyethylene collectors are inexpensive. These low-cost collectors, however, require an annual expenditure for labor and materials to replace the polyethylene.

A solar collector, as well as the storage and distribution system, may be placed within the greenhouse walls (Liu and Carlson 1976). Placing the collectors inside would decrease their heat loss and increase their efficiency (Figure 20), but may reduce the usable space. Tall-growing crops, such as tomatoes, may shade the collector, depending on the angle of the sun and the location of the crops.

Once the air or water is warmed by solar energy, the challenge is to store that heat energy for night use. Heat is transferred from a solar collector to the storage area by water or air. Usually water storage is used with water collectors, and gravel or rockbed storage with air collectors. The volume of water needed to store a given amount of heat is approximately one-third of the volume of gravel or rock; however, because the water systems present problems of corrosion, they require more maintenance than gravel or rockbed storage. Therefore, as gravel bed heat storage techniques improve, air systems will become increasingly popular.

A major drawback to using the greenhouse itself as a collector is the lack of an economically feasible design for storing low-grade heat for night use. In the mid-1970's, research at Cornell University centered around two techniques for storing heat:

Table 14. Performance of a solar collector made of polyethylene (Roberts et al. 1976)

TIME				VARIABLE		
	Flow rate liters/min.	Temp. in°C	Temp. out°C	Insolation KCal-hr/m^2 Available	Collected	Efficiency %
10:00	54	25	28	648	404	62
11:45	54	36	38	718	404	56
13:45	54	40	43	564	282	60

earth storage and rock storage. The earth-air concept was examined for both long-term and short-term storage (Price et al. 1976). For long-term storage, the large block of soil under the greenhouse would be warmed during the summer and thus become a source of heat energy into the late fall and early winter. This system has the potential advantages of not requiring obstructions within the house, and improved plant growth, especially where plants are grown on the warmed ground surface rather than on benches.

If there are benches in the greenhouse, the rock storage may be located under each bench. During the day, the rock would approach the temperatures maintained in the greenhouse. At night, air would be forced through the rock units, the number of units depending on the heating requirements of the greenhouse. Such systems would need auxiliary heat during periods of cold, cloudy weather. To date, neither earth nor rock systems are utilized in commercial production in the United States. In Japan, when winter temperatures are not excessively cold, the earth storage system is used. There the system blows warm air into underground pipes during the day, storing the excess heat in the soil. During the night, the air is warmed when it passes through the underground pipes. Crops having low temperature requirements, such as lettuce, usually require little or no supplemental heating.

Storing the collected heat from solar collectors in water or in a rockbed outside the greenhouse is expensive due to the cost of insulating the water storage tanks or rock beds. Early in 1970, research was initiated to create a system of storing heat in the floor or under the benches within a greenhouse, in a manner

Figure 20. A proposed new concept for maximizing solar energy use in a greenhouse

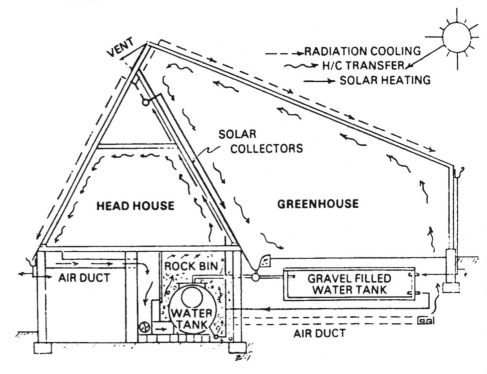

Source: (Liu and Carlson 1976)

different from those systems being studied at Cornell. At Rutgers University, storage systems have been designed into a porous concrete floor within the greenhouse (Roberts et al. 1976). The porous concrete, with a rock aggregate mass underneath, provides space for storing 14 liters of water per .028 m^3 and offers a solid surface on which to walk (Figure 21). Insulation might be required only in those areas where the water table is within 3 m of the floor surface. Warm water from the collector would circulate to the floor and return through the collector during periods of sunny weather.

At the University of Florida, another system was designed as an under bench water storage/heat exchange (Mears and Baird

Figure 21. Cross section of porous concrete floor used for storage of warm water for heating a greenhouse

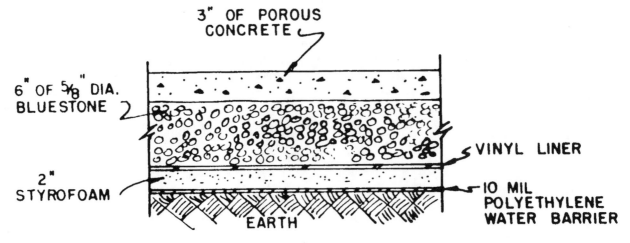

Source: Roberts et al. 1976

Figure 22. Cut-away view of under bench water storage/heat exchanger

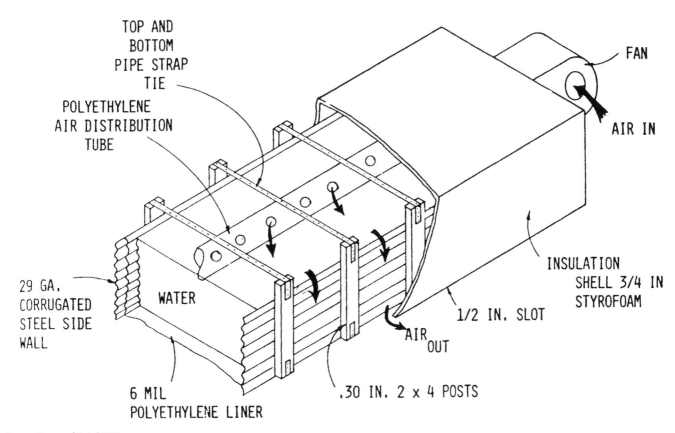

TOP AND
BOTTOM
PIPE STRAP
TIE

POLYETHYLENE
AIR DISTRIBUTION
TUBE

FAN

AIR IN

29 GA.
CORRUGATED
STEEL SIDE
WALL

WATER

INSULATION
SHELL 3/4 IN
STYROFOAM

1/2 IN. SLOT

AIR OUT

6 MIL
POLYETHYLENE LINER

.30 IN. 2 x 4 POSTS

Source: Mears and Baird 1976.

1976). If it is a part of the storage structure, the heat exchanger reduces costs as well as equipment requirements. The entire system must fit under the greenhouse benches and not interfere with normal cultural operations.

The basic components of the Florida system are a plastic water bag, which runs beneath the bench for its entire length, and an insulated arch over this bag to provide an air plenum. A fan with a thermostat control blows greenhouse air to be heated into the space between the arch and the water bag. The air absorbs heat and then exits under a slot the length of the bottom of the arch (Figure 22).

Warm water from a collector would flow into the plastic water bag under the bench and would circulate continuously through the collector and storage tank during periods of sunny weather.

Geothermal Energy. In the U.S. there is increasing interest in the use of geothermal energy for heating greenhouses. Several installations in the western U.S. now utilize this energy source. In Iceland, geothermal energy is much more extensively used as hot water for heating. In Hveragerdi, Iceland, approximately 12 acres of greenhouses are currently utilizing this natural heat, and the Icelandic Government is considering the installation of as many as 80 acres of glasshouses in this region for vegetables and ornamental plant production.

No doubt geothermal installations present abundant opportunities for creative environmental engineering (Axtmann and Peck 1976). As may be expected, this promising energy source is not without its problems, especially those of corrosion and sealing. While these problems vary with each site, attention must be paid to the amounts of gaseous effluents produced by a geothermal well, such as hydrogen sulfide, mercury, radon, ammonia, and H_3BO_3, as well as aqueous emissions containing heavy metals. These, plus the possible complications of silica deposition in the equipment and the disposal of the used thermal fluids, must be carefully assessed before such an energy system is designed into a controlled-environment agriculture installation. The environmental impact of the geothermal system also should be studied.

Waste Heat. A major alternative heat source for controlled-environment agriculture is the rejected heat from large industrial units, such as thermal power electrical generating stations. In 1976, Skaggs et al reported that about 2 kwh of energy were rejected via condenser cooling water for every kwh of electrical power produced. In 1975, the annual quantity of waste heat available in the U.S. was more than 10^{16} BTU, equivalent to 1.6 billion barrels of fuel oil (Madewell et al. 1975). This number represents slightly less than 20 percent of the annual energy used in the U.S. in 1971. Approximately 95 percent of the water used in cooling electric generator condensors was returned to streams or estuaries. One nuclear power plant composed of 3 generating units, each having a generating capacity of 1,270 megawatts, rejects at full load approximately 10×10^9 BTU/hr. This is sufficient net energy to heat almost 4,000 ha. of conventional greenhouses.

It is difficult to economically utilize the waste heat energy for heating greenhouses because of the low temperature of the rejected water, although systems have been tested to extract the heat from the warm water of large industrial units.

In 1977, a one-half acre greenhouse was constructed at the TVA Browns Ferry Nuclear Plant in Alabama. A portion of this greenhouse was heated with condenser water from the power plant. The design of the system incorporated heating concepts conceived and tested by the Oak Ridge National Laboratory, Tennessee Valley Authority and the Environmental Research Laboratory of the University of Arizona. The cost to connect and install the hot water delivery system from the power plant to the greenhouse was extremely expensive and therefore the system was not cost effective for application in the greenhouse industry.

Similar projects have been tested in France and England, as well as in the United States and other countries. With rare exceptions, none of these programs has been expanded into commercial use, due to the high cost of connecting the greenhouses to a power plant and the unreliability of a heat source from a power plant, especially during times of plant repair and maintenance.

Atmosphere Control. The gaseous makeup of the greenhouse environment is important if crop yields are to be maximized. Carbon dioxide must be maintained at ambient levels, or greater, and toxic gas build-up must be prevented by providing adequate combustion air for the heating system.

Carbon Dioxide Enrichment. Photosynthesis — the production of carbohydrates from carbon dioxide (CO_2) and water in the presence of chlorophyll, using light energy — is basic to plant growth and reproduction. The rate of photosynthesis is governed by available fertilizer elements, including water, CO_2, light, and temperature (Figure 23).

Under normal conditions, CO_2 exists as a gas in the atmosphere at a level slightly over 0.03 percent or 330 parts per million (ppm). During the day, when photosynthesis occurs under natural light conditions, the plants in a greenhouse may draw down or reduce the level of CO_2 to below 200 ppm. Under these circumstances, CO_2 levels are increased by infiltration into the greenhouse, or if ventilation is provided, outside air may be brought in to maintain the CO_2 at ambient levels. If the level of CO_2 is maintained at less than ambient levels, CO_2 may become the factor limiting for plant growth.

In cold climates, maintaining ambient levels of CO_2 by providing ventilation may be uneconomical, due to the necessity of heating the incoming air in order to maintain proper growing temperatures. In such regions growers commonly enrich the greenhouse with added CO_2. The exact CO_2 level needed for a given crop will vary, since it must be correlated with other variables in greenhouse production such as light, temperature, nutrient levels, cultivar and degree of maturity. Most crops will respond favorably to CO_2 at 1,000 to 1,200 ppm.

In the early 1960's, pure CO_2 was supplied from cylinders, dry ice, or tanks of the low pressure liquid. Some of these methods of application may still be used. More prevalent today is the use of combustion units of varying degrees of refinement, employing fuel oils, propane or natural gas. The burners can

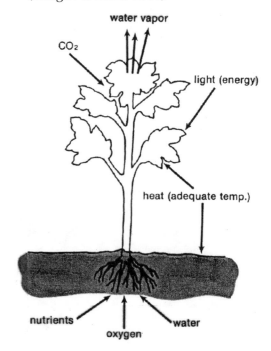

Figure 23. Conditions necessary for plant growth (Badger & Poule 1979)

either be within or outside the greenhouse. In recent years, the Dutch have developed a new concept in CO_2 enrichment using a centralized CO_2 generator.

The records of costs and returns have demonstrated an increase in yields and financial returns, often far beyond the costs of CO_2. But CO_2 enrichment is not always practical. In regions where ventilation is required, even on many winter days, such as in the desert areas, added CO_2 is lost through the ventilation systems. Because it is not contained within the greenhouse, added CO_2 does not affect plant growth. In such situations, CO_2 enrichment is not advised and does not show any economic advantage.

If ventilation is not needed for any period of time, whether the greenhouse is located in a desert region or in cooler climates where CO_2 enrichment is not used, outside air should be allowed into the greenhouse periodically either by a ventilation fan drawing fresh air into the greenhouse unit or by opening the vents slightly to allow air exchange that will maintain levels of ambient CO_2 (330 ppm). A forced-air distribution system, using perforated plastic ducts separate from or in combination with the heat distribution system, should be operated continuously in order to distribute the CO_2 evenly within the plant canopy. This movement of air will prevent the CO_2 from becoming lower than ambient within the canopy.

Above a concentration level of 5,000 ppm, CO_2 is hazardous to workers. Most plants also have a maximum tolerance level, depending on the cultivar.

Contaminate Gases. Proper installation and maintenance is crucial to prevent contaminate gases in a greenhouse. Inadequate heating equipment and operation will endanger not only the health of the plants but also those persons working within the greenhouse. Annual maintenance inspections and

Figure 24. Provision of combustion air and sufficient chimney to prevent contaminated gasses

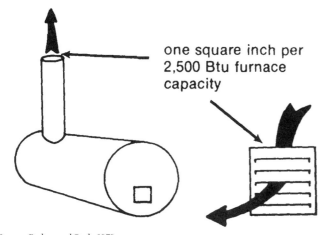

one square inch per 2,500 Btu furnace capacity

Source: Badger and Poole 1979.

adjustments are a minimal requirement for efficient and safe operation.

Inadequate combustion air within the greenhouse prevents complete combustion and proper venting of combustion gases. The products of incomplete combustion may be discharged into the greenhouse from the draft diverter on the heater. The unburned gas, carbon monoxide, and other chemical gases, will create an unsafe environment for plants and workers.

To prevent hazards, it is important to instill louvered fresh air intakes from outside, near the heating unit. These intakes will assure complete combustion of the fuel within the furnace (Figure 24).

This is extremely important in greenhouses that are of tight construction, especially those covered with plastic film. Allow one square inch of louvered air inlet area from outside for each 2,500 Btu of furnace capacity. Louvers made of differing materials require different amounts of free area: wooden louvers usually have only 20 percent free area; metal louvers commonly have 60 percent free area when open.

Incomplete combustion also causes soot buildup within the furnace, which reduces heat transfer and lowers heater efficiency.

Chimney air leaks chill the gases and reduce the draft, and pollution from furnaces and chimney leaks may injure plants.

Computer/Data Acquisition Systems. Today, computer control systems are common in greenhouse installation throughout Europe, Japan, and the United States. Computer systems can provide fully-integrated control of temperature, humidity, irrigation and fertilization, carbon dioxide, light and shade levels for virtually any size growing facility. Precise control over a growing operation enables growers to realize savings of 15 - 50 percent for energy, water, chemical, and pesticide applications. Computer controls normally result in greater plant consistency, on-schedule production, higher overall plant quality, and environmental purity.

A computer can control hundreds of devices within a greenhouse (vents, heaters, fans, hot water mixing valves, irrigation valves, curtains, lights, etc.) by utilizing dozens of input parameters, such as outside and inside temperatures, humidity, outside wind direction and velocity, carbon dioxide levels and even the time of day or night.

Computer systems interrogate all sensors, evaluate all conditions, and send appropriate commands every minute to each piece of equipment in the greenhouse range thus maintaining ideal conditions in each of the various independent greenhouse zones defined by the grower.

Computers collect and log data provided by greenhouse production managers. A computer can keep track of all relevant information, such as temperature, humidity, CO_2, light levels, etc. It dates and time tags the information and stores it for current or later use. Such a data acquisition system will enable the grower to gain a comprehensive understanding of all factors affecting the quality and timeliness of the product. A computer will produce graphs of past and current environmental conditions both inside and outside the greenhouse complex. Using a data printout option, growers can produce reports and summaries of environmental conditions such as temperature, humidity, and the CO_2 status for a given day, or over a longer period of time.

Scientists are currently developing plant growing models in which computers actually make decisions for the greenhouse growers. These "artificial intelligence" systems integrate the latest knowledge about greenhouse growing theory, actual management practices, and environmental conditions inside and outside the greenhouse. The computer will be taught to assess all the variables, make a decision, and give instructions for application. The decisions made by the computer in climate control can provide the grower 24 hour-a-day assistance in the management and production of greenhouse crops. A system can be so reliable that, even if it should fail, it will not only call the grower on the telephone but will also turn key components over to local control.

Unfortunately, such computer systems are expensive and mainly limited to large greenhouse facilities operating year-round. The crops grown are usually of high value and are those that respond to precise control over the environment. Computers are not economically feasible for protected agriculture situations that are seasonal: the added costs outweigh the economic benefits unless used throughout the year.

Despite the attraction of computer systems, it is well to remember that the success of any production system is totally dependent on the farmer's knowledge and his management skills. Computers only assist by adding precision to these skills. A computer is only as effective as the person who feeds it the data.

As computer costs continue to decrease and as farmers become computer literate, computers will become increasingly popular in protected agriculture. In developing countries, where farmers lack formal education, financial resources, and the skill to operate computers, the utilization of these systems in protected agriculture is remote.

PART THREE

PRODUCTION ASPECTS

CHAPTERS 5 - 10

5

GROWING SYSTEMS IN GREENHOUSES

There are numerous greenhouse production systems. Many are applicable only in some regions of the world. Those systems using indigenous materials are likely to be more economical than those requiring imported materials. Systems of production must be carefully evaluated before any is selected. Plastic covered greenhouses are the most popular structures used in protected agriculture.

TISSUE CULTURE

Tissue culture is a relatively new method of plant propagation. It is available to growers for two major purposes:

(l) mass production and (2) to establish and/or maintain "virus free" stock (Kyte 1987). With one small piece of plant, numerous replicas, or clones, can be produced. The term "cloning" is synonymous with "tissue culture", "micropropagation", or growing *"in vitro."* *In vitro* (Latin, "in glass") means separate from the whole individual. Vegetative reproduction, or "cloning", can be accomplished by taking cuttings, making divisions, by layering, grafting, or by tissue culture.

Tissue culture begins in a laboratory, which consists of a media preparation room, a transfer room, and a culture growing room. The plants are started in the tiniest of greenhouses — a test tube — in a sterilized growing medium consisting mainly of agar (a polysaccharide gel derived from a certain algae), fertilizer elements, vitamins, and growth regulators. Once the plant tissue has developed into a plantlet, it is transferred to a soilless media: any combination of peat moss, sand, perlite, pumice, or vermiculite. The plantlets are moved to a greenhouse and begin the process of acclimatization. Once acclimatized, or hardened off, they are placed under normal greenhouse conditions to growing out.

The genetic makeup of plants which are cloned or vegetatively reproduced is identical to the parent. They are therefore fundamentally different from plants propagated from seed: seeds resulting from sexual reproduction have their own unique genetic makeup, a mixture from both parents; the overall genetic make-up of the seed is different from that of either parent, and differs from one seed to another.

In nature, vegetative reproduction occurs naturally in multiplication of stems. True bulbs such as tulips and daffodils are modified stems. Another type of underground stem is the corm which is complete with nodes, internodes, and lateral buds, but does not have scales, as do true bulbs. Gladiolus and crocus are corms. Other modified stems are tubers, such as potato; rhizomes, which are common for most perennial grasses, and runners, or stolons, as with strawberries. Stolons are above ground vegetative reproductive stems, while rhizomes are below ground.

Tissue culture laboratories are becoming increasingly popular as a method of plant propagation. This method allows for mass propagation of plants and establishment of virus-free plant stock.

Roots and leaves can also serve as vegetative reproductive organs. Examples of plants that reproduce by roots are sweet potatoes and dahlias, while those reproducing from leaves are Rex *begonia* and African violets.

In the horticultural industry, tissue culture practices usually involve the use of meristematic tissue.

Meristematic cells are located at the top of stems and roots, in leaf axils, in stems as cambium, on leaf margins, and in callus tissue (Kyte 1987). The tissue is composed of cells which have not yet been genetically programmed for their ultimate development.

The factors involved in meristematic tissue culture have been well-outlined by Kyte (1987):

"Under the influence of genetic make-up, location, light, temperature, nutrients, hormones, and probably other factors, meristematic cells differentiate into leaves, stems, roots, and other organs and tissues in an organized fashion.

Some differentiated cells (usually parenchyma cells) have the capability of reverting to a meristematic or dedifferentiated state to initiate growth of new and different tissue. Dedifferentiated cells often account for adventitious growth. Adventitious growth refers to the growth of new shoots, buds, roots, or leaves from unusual locations. Examples are aerial roots, buds from roots, plantlets from leaves, shoots and roots from callus of cuttings.

Dedifferentiated cells can also create callus. Callus is usually defined as a mass of undifferentiated cells, or parenchyma cells proliferating in response to wounding, as appears at the union of a graft, or at the base of cuttings. In tissue culture vocabulary callus is defined as an unorganized, proliferating mass of cells. Some of the cells may be differentiated, so the mass may contain embryoids (embryo-like structures capable of becoming normal plants), or the mass may contain shoot or root primordia, or there may be cells with an abnormal number of chromosomes. For example, many asparagus plants differentiating from the callus culture may be tetraploid (double the normal chromosome number in vegetative cells), but plants cultured from shoot tips, without a callus stage, do not show this variability.

Whenever cells divide there is the possibility of genetic variability. There are two kinds of plant cell division, meiotic and mitotic. Meiotic division relates only to sexually reproductive cells. Mitotic (somatic, or vegetative) cell division is the division of a vegetative cell to produce two cells each of which usually has the same number of chromosomes as the original cell. If a mutation - a change in chromosomes -occurs during mitosis, it is carried in all future divisions. Some somatic mutations go unnoticed because future cells are not affected; if a mutation for aberrant flowers occurs in a leaf bud, there will be no flowers to exhibit the mutation. Most mutations produce undesired effects: misshapen fruit or abnormal shoots, for example. Growers discard such plants when they appear. Occasionally mutations are desirable so scientists induce mutations with chemicals and radiation in a search for better plants.

Plant diseases transmitted from parent to offspring can often be eliminated through tissue culture procedures. External infestations, such as bacteria, fungi, and insects, may be removed when cleaning the tissue to be cultured. Many internal viruses are eliminated by using the apical meristem as the explant. Apical meristem, the new, undifferentiated tissue at the apex, or very tip of a shoot, is usually virus-free in diseased plants, since the new cells normally grow more rapidly than the viruses, which lag behind in the older tissue. These few cells that make up the microscopic portion of the apical meristem are removed from the plant and placed in culture; they will grow and produce healthy, disease-free plants. This technique, known as meristem culture, is used extensively both in research and commercial laboratories to generate virus-free plant material from lilies, dahlias, carnations, citrus, potatoes, and berries, particularly strawberries (Kyte 1987).

Tissue culture, or micropropagation, proceeds today through application of many advanced techniques, such as somatic embryogenesis, protoplast isolation, and protoplast fusion. Along with haploid cultures, which are derived from pollen by way of anther culture, these techniques are mainly used in research facilities where genetic engineering is underway. While these procedures are not commonly used by plant growers, they are potentially important to the future of food production: they may produce plants resistant to frost, drought, salinity, insect pests and diseases, control nitrogen fixation in grasses.

TRANSPLANT AND PLUG MIXTURE PRODUCTION

Excellent crops of high yield begin with seed that is disease-free and of high quality. Since good seed is probably the biggest bargain in plant growing, it is better to pay a little more to a seed supplier of known dependability than to get a supposed bargain from an unknown company.

Plant Growing Media. A good plant growing medium should be friable, moderately fertile, of good aeration, and well-drained, yet have sufficient water-holding capacity to prevent excessive drying. It should also be free of weed seeds, and diseases, economical, and relatively easy to obtain. Plant growing media vary from just topsoil to combinations of topsoil, sand, peat moss, vermiculite, perlite, rock wool, rice hulls, compost, or other additives. Plants may be grown in completely artificial media.

Due to problems of diseases, weed seeds, drainage, aeration, and inconsistency in physical conditions, pure topsoil as a seeding and transplant media is not recommended. If pure soil is used, it must be a sandy loam with high organic matter content. Adding peat moss or a combination of peat moss and sand to soil usually improves the water-holding capacity and physical condition of the medium. Most soil mixtures of this nature will need additions of limestone, superphosphate and a 5-10-5 fertilizer mix. The rate of application would be 1.5 to 2.5 kg. of 20% superphosphate, 4.5 kg. ground limestone and 1-1.5 kg. of 5-10-5 fertilizer per cubic meter.

Any plant growing media containing soil should be fumigated or steamed to destroy weed seeds and disease-causing organisms. Fumigation with methyl bromide is effective if directions are followed but this fumigant is highly toxic. A 2% mix of chloropicrin or tear gas added to methyl bromide should be used as the chloropicrin serves as a warning agent in case of gas leakage in an inhabited area. In some countries, such use of fumigant gases is illegal for sterilization of plant growing media; therefore it is important to check the local and federal regulations before use.

Steam is an effective method of treatment; however, because the cost of fuel to operate the steaming apparatus has become expensive, steaming is currently less attractive for use in greenhouse operations. If steam is used, the soil mix may be steamed in concrete bins, dump trucks, or other suitable containers. It should be covered with a material that will contain the steam. For effective disease and weed seed control, the soil should reach 82°C and remain at that temperature for 30 minutes. Over-steaming should be avoided: excessive heat may result in the breakdown of organic matter and buildup of materials in the soil to levels which might cause plant injury. Soil should not be used within two weeks after steaming.

Artificial media and mixtures are finding more favor with growers because good topsoil is increasingly rare, regulations

Rockwool growing blocks are commonly used in the production of cucumber transplants and other greenhouse vegetable crops.

ture is lightweight and, when thoroughly mixed, will produce uniform, high-quality plants. Mix the ingredients on a clean floor or in a mechanical mixer. The floor area and all the tools should be washed with a disinfecting solution such as one part Clorox bleach to ten parts water. (The active ingredient in Clorox is sodium hypochlorite, also common to other brands of bleach.) It is easier to handle a peat-lite mix if the peat moss is moistened prior to or during mixing. Peat moss may be difficult to moisten, especially if very dry. Peat will take up moisture if the water is hot or if the water is added in combination with a wetting agent (Agra-Grow granular, 350 g/cu.m.). After mixing thoroughly, use immediately or store under a plastic sheet to retain the moisture. A moist medium absorbs water more uniformly and readily than a dry one. Peat-lite mixes have been stored satisfactorily up to one year.

place increasing restrictions on the use of fumigants, and fuel costs for steam sterilization continue to escalate. Artificial media should be sterile, of known nutrient concentration, easily stored and light in weight. Mixtures of peat and perlite or peat and vermiculite have these attributes. In some cases, vermiculite alone has proven a satisfactory medium for certain directly-seeded vegetables and flowers.

In tropical regions of the world, where peat is not indigenous and very expensive, other materials such as shredded coconut husk or rice hulls are often used in combination with vermiculite and sand. Rice hulls should be carbonized with high temperature in order to prevent rapid breakdown by microorganisms.

In desert regions, where peat is also expensive, vermiculite mixed with sand will provide an excellent plant growing medium.

Considerable research has been conducted at two state universities in the United States, the University of California and Cornell University, on artificial plant-growing mixes (Schales and Massey 1965). The growing mixtures developed by these institutions are far superior to soil in aeration and drainage characteristics.

The University of California (U.C.) mixture is basically a variable formula of sand and peat moss with several possible fertilizer combinations. A mixture that is popular is the following:

U.C. Mix	Amt. per cubic meter
Fine sand (0.5-0.05 mm)	0.5 cu. m.
Shredded peat moss	0.5 cu. m.
Potassium nitrate	45 g.
Potassium sulfate	45 g.
Dolomitic limestone	4.5 kg.
Calcium limestone	1.5 kg.
20% superphosphate	1.5 kg.

The Cornell "Peat Lite" mixes are made up with peat moss and vermiculite or perlite. This peat moss and vermiculite mix-

Cornell Peat-Lite Mix	Amt. per cubic meter
Shredded peat moss	0.5 cu. m.
No. 2, 3, or 4 horticultural grade vermiculite (or perlite)	0.5 cu. m.
Limestone, dolomitic preferred	3.00 kg.
Superphosphate (20% P_2O_5)	1.20 kg.
Fertilizer (5-10-5)	3.00 kg.

If another medium is substituted for peat moss and peat is not used in the U. C. mix or the Cornell Peat-Lite mix, limestone is not required, as lime is used mainly to raise the pH of the mixture.

In those cases where the plant is physically pulled by the stem from the plant container, the root ball is more likely not to break apart if peat is in the artificial growing media. This happens because the root growth within the peat moss fibers tends to hold the roots and media together as a unit. With mixtures not containing peat moss the root unit will fall apart more easily; therefore care must be taken not to damage the root system at the time of transplanting.

Seed Germination. Seed can be sown directly into individual plant containers containing growing media, into grow cubes of rock wool or Oasis root cubes, or as rows into wooden or plastic flats containing a plant growing mix or pure vermiculite. If flats are to be reused, they should be thoroughly cleaned and rinsed in Clorox (sodium hypochlorite).

Fertilization. It is recommended that no supplemental fertilizer be added to the mix or the vermiculite until the seed leaves (cotyledons) are fully expanded and the true leaves are beginning to unfold . Fertilization should be in liquid form with the fertilizer, 20-20-20, mixed at the rate of 0.35 to 0.45 kg. per 200 liters of water and used over an area of 20 square meters. The length of time before the seedlings are ready to be spotted out into individual containers will depend on the species of plant grown.

The plants will need to be fertilized either with a very dilute fertilizer solution with each watering (for example, 110-175 grams of 20-20-20 per 200 liters of water) or fertilized approximately every two weeks at a rate of 500-700 grams of 20-20-20, or similar fertilizer, to each 200 liters of water. In the latter procedure, to prevent fertilizer burn of the leaves, the leaves will need to be rinsed with pure water after each liquid feeding. Fertilizers used for liquid feeding should be 100 percent water soluble.

Figure 25. Types of trays used in the production of transplants

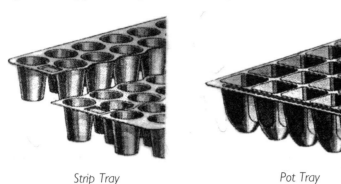

Strip Tray   *Pot Tray*   *Tube Tray*

Containers. Growing containers come in many forms. Individual containers or growing blocks may be made of paper, plastic, clay, peat moss, styrofoam, rockwool, or a synthetic foam. Growing trays, plug flats, cavity seedling trays, styrofoam planter flats, and multipot plug trays, and others are all multiple seedling trays common in today's industry. In Figure 25, several types of trays used in the transplant industry are illustrated.

Mechanical Aids. There are many types of mechanical aids used in transplant production. Today, simple to operate and inexpensive mechanical seeders are available as illustrated.

Once the flats are seeded they may be manually transported to the greenhouse or transplanted by monorail trolley conveyors. The same system of transport is used in taking the trays out of the greenhouse in preparation for shipment to market.

Watering. There are many methods of watering transplants. Where plants are hand-watered, round sprinkler heads deliver a greater amount of water and do less damage than fan-shaped sprinkler heads.

Each watering should soak the soil thoroughly. Water should be applied uniformly over all plants to avoid uneven growth. Plants should be watered early enough in the day to permit foliage to dry before dark.

Overhead irrigation systems save considerable time but have difficulty obtaining completely uniform water distribution. Today, it is common for growers to use programmable travelling irrigation booms.

GREENHOUSE VEGETABLES

Growing greenhouse vegetables is one of the most exacting and intensive forms of all agriculture enterprises. In the controlled environment of a greenhouse, high yields of excellent quality vegetables can be produced.

Soil. The standard method of growing greenhouse vegetables throughout the world is in soil. A successful grower who grows

Mechanical seeders are simple and inexpensive to operate. (Photo courtesy Growing Systems, Inc. Milwaukee, Wisconsin).

Monorail trolley conveyors and movable benches quickly and efficiently move stock in and out of greenhouses. (Photo Courtesy of Growing systems, Inc. Milwaukee, Wisconsin).

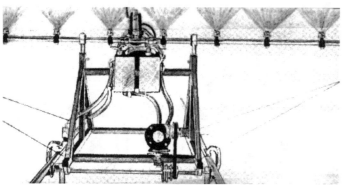

Fully programmable travelling irrigation booms for automatic uniform watering, fertilizing, etc. (Photo courtesy of Growing Systems, Inc. Milwaukee, Wisconsin)

in soil usually has a good knowledge of horticulture, soils, plant pathology, entomology, and plant physiology, as well as the engineering capability to provide an environment best suited for plant growth. Many persons who establish a greenhouse operation fail because they lack the education and training in one or more of the above disciplines.

A major problem in growing crops in soil are soil-borne diseases. Growing plants continuously, without crop rotation or interruption in production, as in open field production during Northern winters, can lead to an excessive build-up of soil pathogens. Because of environmental and health restrictions, there is currently a lack of soil fumigants available for greenhouse use This problem, added to the high cost of fuel to steam sterilize, is focusing attention on methods of hydroponic controlled environment agriculture (CEA).

HYDROPONIC/SOILLESS CULTURE

Hydroponic culture is possibly the most intensive method of crop production in today's agricultural industry. In combination with greenhouses or protective covers, it is high technology and capital intensive. It is also highly productive, conservative of water and land, and protective of the environment. Yet, for most of its employees, hydroponic culture requires only basic agricultural skills. Since regulating the aerial and root environment is a major concern in such agricultural systems, production takes place inside enclosures designed to control air and root temperatures, light, water, plant nutrition, and adverse climate.

During the last 12 years, there has been increasing interest in hydroponics or soilless techniques for producing greenhouse horticultural crops. The future growth of hydroponics depends greatly on the development of production systems that are cost competitive with open field agriculture.

There are many types of hydroponics systems, as well as many designs for greenhouse structures and many methods of control of the environment. Not every system is cost effective in each location. While the techniques of hydroponic culture in the tropics may be quite similar to those used in temperate regions, greenhouse structures themselves and methods of environmental control can differ greatly.

Hydroponics is a technology for growing plants in nutrient solutions (water and fertilizers) with or without the use of an artificial medium (e.g. sand, gravel, vermiculite, rockwool, peat moss, sawdust) to provide mechanical support. Liquid hydroponic systems have no other supporting medium for the plant roots; aggregate systems have a solid medium of support. Hydroponic systems are further categorized as open (i.e., once the nutrient solution is delivered to the plant roots, it is not reused) or closed (i.e., surplus solution is recovered, replenished, and recycled).

Some regional growers, agencies, and publications persist in confining the definition of hydroponics to liquid systems only. This exclusion of aggregate hydroponics serves to blur statistical data and may lead to underestimation of the extent of the technology and its economic implications.

Virtually all hydroponic systems in temperate regions of the world are enclosed in greenhouse-type structures to provide temperature control, reduce evaporative water loss, reduce disease and pest infestations, and protect crops against the elements of weather, such as wind and rain. The latter considerations are especially valid in tropical regions. While hydroponic and controlled environmental agriculture (CEA) are not synonymous, CEA usually accompanies hydroponics. Their potentials and problems are inextricable.

The principle advantages of hydroponic CEA include high-density maximum crop yield, crop production where no suitable soil exists, a virtual indifference to ambient temperature and seasonality, more efficient use of water and fertilizers, minimal use of land area, and suitability for mechanization and disease control. A major advantage of hydroponics, as compared with growth of plants in soil, is the isolation of the crop from the underlying soil, which often has problems of disease, salinity, poor structure and drainage. The costly and time-consuming tasks of soil sterilization and cultivation are unnecessary in hydroponic systems and a rapid turnaround of crops is readily achieved.

Hydroponics offers a means of control over soil-borne diseases and pests, which is especially desirable in the tropics, where infestations are a major concern. Most temperate regions have climatic changes, such as cold winters, to break the life cycles of many pests. In the tropics, this life cycle continues uninterrupted, as does the threat of infestation. Unfortunately, less is known about many of the diseases that occur in the tropics than those in temperate regions. In comparing three major food crops grown in the tropics and temperate regions of the world, the incidence of disease is much greater in the tropics, as illustrated in Table 15.

Table 15. Number of crop diseases

CROPS	TEMPERATE	TROPICS
Rice	54	500-600
Corn	85	125
Beans	52	250-280

Source: (Wittwer 1981)

The principle disadvantages of hydroponics, relative to conventional open-field agriculture (OFA), are the high costs of capital and energy inputs, and the high degree of management skills required for successful production. Capital costs may be especially excessive if the structures are artificially heated and evaporatively cooled by fan and pad systems, systems of environmental control which are not always needed in the tropics. Workers must be highly competent in plant science and engineering skills. Because of its significantly higher costs, successful applications of hydroponic technology are limited to crops of high economic value to specific regions, and often confined to specific times of the year, when comparable OFA is not feasible.

Because with CEA, capital costs are so much higher than for OFA, it is economically rational to grow only a few food crops. Field crops are totally inappropriate. A decade ago, it was calculated that the highest market prices ever paid would have to increase by a factor of five for hydroponic agronomy to break even. Since then, CEA costs have more than doubled, while crop commodity prices have remained constant. Indeed, in the United States, open-field agronomic crops are usually in surplus, and a significant percentage of the available cropland is deliberately idled. Repeated pricing studies have shown that only high-quality, garden type vegetables - tomatoes, cucumbers, and specialty lettuce - can provide break even or better revenues in hydroponic systems. These are, in fact, virtually the only hydroponic CEA food crops grown today in the United States. In Europe and Japan, these vegetables, and eggplant, peppers, melons, strawberries, and herbs, are grown commercially in hydroponic systems.

Liquid (nonaggregate) Hydroponic Systems. By their nature, liquid systems are closed systems in which the plant roots are directly exposed to the nutrient solution, with no other growing medium, and the solution is reused.

Nutrient Film Technique (NFT). The nutrient film technique was developed during the late 1960's by Dr. Allan Cooper at the Glasshouse Crops Research Institute in Littlehampton, England (Winsor et al. 1979); a number of subsequent refinements have been developed at the same institution (Graves 1983). NFT has given rise to several modified systems and appears today to be the most rapidly evolving type of liquid hydroponic system.

In a nutrient film system, a thin film of nutrient solution flows through plastic lined channels, which contain the plant roots. The walls of the channels are flexible; this permits them to be drawn together around the base of each plant, excluding light and preventing evaporation. The main features of a nutrient film system are illustrated in Figure 26.

Nutrient solution is pumped to the higher end of each channel and flows by gravity past the plant roots to catchment pipes and a sump. The solution is monitored for replenishment of salts and water before it is recycled. Capillary material in the channel prevents young plants from drying out, and the roots soon grow into a tangled mat.

A principle advantage of the NFT system in comparison with others is that it requires much less nutrient solution. It is therefore easier to heat the solution during winter months, to obtain optimum temperatures for root growth, and to cool it during hot summers in arid or tropical regions, thereby avoiding the bolting of lettuce and other undesirable plant responses. Reduced volumes are also easier to work with if it is necessary to treat the nutrient solution for disease control. A complete description on the design and operation of an NFT system is published in Horticultural Review, volume 7, pages 1-44. Following are additional comments important to the design of NFT systems for the tropics.

The channels should not be greater than 15-20 m in length. In a level greenhouse, longer runs could restrict the height available for plant growth, since the slope of the channel usually has a drop of 1 in 50 to 1 in 75. Longer runs and/or channels, with less slope, may accentuate problems of poor solution aeration.

To assure good aeration, the nutrient solution could be introduced into channels at two or three points along the length. The flow of nutrient solution into each channel should be 2-3 liters per minute, depending on the oxygen content of the solution. The maximum temperature of the nutrient is 30°C. Temperatures above 30°C will adversely affect the amount of dissolved oxygen in the solution. There should be approximately 5 ppm or more, especially in that nutrient solution flowing

Figure 26. Main features of a nutrient film hydroponic system

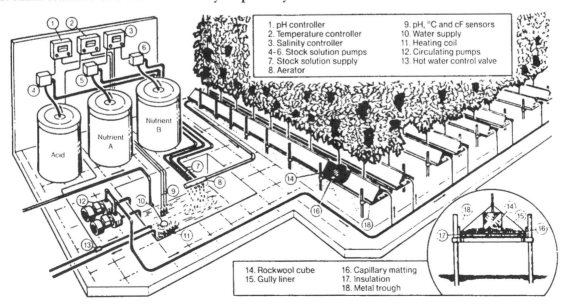

1. pH controller
2. Temperature controller
3. Salinity controller
4-6. Stock solution pumps
7. Stock solution supply
8. Aerator
9. pH, °C and cF sensors
10. Water supply
11. Heating coil
12. Circulating pumps
13. Hot water control valve

14. Rockwool cube
15. Gully liner
16. Capillary matting
17. Insulation
18. Metal trough

Tomato production using the nutrient film technique, a growing method developed in England.

of the concrete was painted with epoxy resin to isolate it from the nutrient solution. In such an installation in Cheshunt, England (Jensen 1982), 4 ha of concrete NFT channels have permitted a high degree of mechanization in planting and harvesting, with resulting lower production costs ($.145/head in summer; $.270/head in winter, higher due to heating costs).

Movable Channels. In a modified NFT system first proposed by Prince *et al.* (1981) for lettuce production, the channels can be spread apart in response to plant growth and size. This variable plant spacing technique maximizes space utilization and leaf interception of radiation. There are limited uses of this system in Canada and the United States.

Movable benches covered with corrugated sheets have been used as lettuce planting troughs in some experiments in the United States and Denmark. Plants are set in the corrugations, through which nutrient solution flows, at close intervals when young, and spread out as more growing space is needed. Benches are movable to allow access to growing areas.

over the root mat in the channel.

In tropical regions it is important that the channels be colored white in order to mitigate the problems of heat build-up from direct sunlight. Normally a white on black plastic is used for the channels. High air temperatures are common in the tropics. An NFT system may permit economical cooling of plant roots, avoiding the more expensive cooling of the entire greenhouse aerial temperature.

In research conducted by Jensen (1985), it was found that root temperatures of lettuce must not exceed 20°C, especially when aerial temperatures are 32-35°C or greater, to avoid the problem of bolting (formation of seed stalk). Cooling the nutrient solution dramatically reduces bolting and lessens the incidence of the fungus *Pythium aphanidermatum*, which also affects the establishment and yield of hydroponic tomato and cucumber crops.

In 1983, the capital cost of an NFT growing system was estimated at $81,000/ha. (not including the cost of construction labor and the greenhouse enclosing the system); the annual operating cost was $22,000/ha (Van Os, 1983) for replacement of plastic troughs and other items. High installation costs and the introduction of rockwool have diminished the popularity of NFT.

Flat sheets of expanded polystyrene, approximately 2 m x 4 m x 2.5 cm, which can be transported on movable benches, have been used for lettuce production by Varley and Burrage (1981). The sheets, framed in wood, are covered with plastic film to form a wide water trough; additional polystyrene sheets drilled with holes for lettuce seedlings ($20/m^2$) are fitted into the troughs. This system has been further simplified in a Dutch application.

Schippers (1982) has suggested that tomato and cucumber production could be maximized in a limited area by mounting channels on casters and arranging them without intermediate pathways. This would increase plant density by approximately

Modified NFT.

Cast Concrete. Covering a greenhouse floor with cast concrete shaped into NFT channels is one type of modified NFT system. Capital investment is higher, but maintenance is reduced compared with a standard system. In a typical installation for year-round lettuce production (Lauder 1977), the concrete was formed into parallel channels 10 cm wide, 2.5 cm deep, and 45 m long, on a slope of 1 to 50. The surface

Lettuce production in a modified NFT system with shallow troughs of cast concrete.

Figure 27. Lettuce production in modified NFT system with movable belts in troughs stacked two high

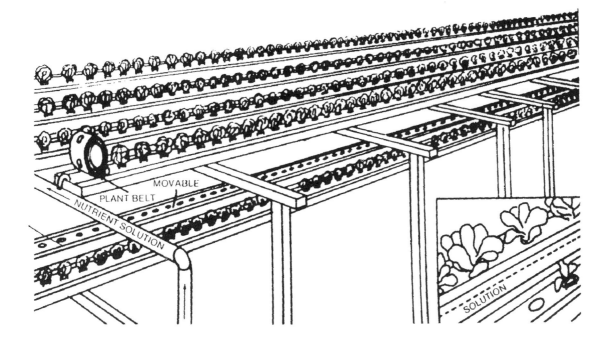

25%. When cultural operations are required, channels may be moved and separated to create a pathway.

Giacomelli developed a design at Rutgers University for planting tomatoes in flexible plastic tubes suspended by a movable system (Giacomelli *et al.* 1982), thus transferring the weight of the rows to greenhouse structural members. Such easily moved rows would permit close spacing of young plants and wider spacing as plants mature, with a potential 25 percent increase in annual yield per unit area of greenhouse space.

Van Os (1983) has described a completely mechanized NFT system to produce cut chrysanthemums: mobile planting tables or "transporters" are moved from room to room for planting, growing, and machine harvesting, with a reduced number of interior pathways. Potter and Sims (1975) have also referred to a system of movable channels for such crops.

Multicropping. In Holland, research engineers at Wageningen are using NFT in a system to multicrop lettuce with CEA tomatoes. The system is normally used to grow tomatoes, but during winter months, additional 25-cm-wide troughs in each bay of the greenhouse are used to produce automated-harvest NFT lettuce. The heads are planted through holes in a flexible plastic material that covers each trough. At harvest, a winching machine pulls the covering material, lettuce and all, up an incline to be rolled up on a spool; as the plastic moves upslope toward the winch, a cutting mechanism severs lettuce heads from roots. The lettuce moves off on a conveyor belt toward the packing station, as the roots are removed on a different conveyor and the plastic cover is slowly wound up on the drum (Jensen 1982).

Similar, if less mechanized, multicropping systems have been developed in England (Starkey 1980). Tomatoes or cucumbers are grown in 35-cm-wide NFT channels during spring, summer, and fall; double rows of lettuce, celery, or Chinese cabbage are grown in the same troughs during winter months.

Pipe Systems. An A-frame system developed by Morgan and Tan (1982) provides for high-density lettuce production. Seedlings are planted in sloping (drop of 1 in 30) plastic tubes, 30 mm in diameter, arranged in horizontal tiers resembling A-frames in end-view. This system, developed for use with Dutch Venlo glasshouses, effectively doubles the usable growing surface and accommodates a plant density of $40/m^2$. Another A-frame system of tiered NFT channels, developed for strawberry production, is reported to facilitate spraying and picking, with quick crop turnaround and reduced labor requirements (White, 1980).

Schippers (1978) developed a vertical pipe system in which small-diameter plastic pipes, 1.3 m long, are suspended by overhead wires above a nutrient solution-collecting channel. Germinated lettuce plants are squeezed into holes (20-28/pipe) in the sides of the tubes, and nutrient solution is pumped into the top of each pipe, to drip down through the tubing and plant roots. A similar system with aluminum irrigation piping has been used in Israel to produce vegetables and flowers. Vertical NFT systems using plastic or aluminum piping are relatively costly, and they have not been widely adopted in commercial operations.

Moving Belts. To maximize lettuce production in a limited area, Whittaker Agri-Systems developed a highly mechanized NFT system (Rogers, 1983) in a 0.8-ha CEA complex in California. Twenty-day-old seedlings are transplanted into movable belts in NFT troughs stacked two high and mechanically harvested after 19-35 days of growth; conveyor belts transport the lettuce

Figure 28. System layout of floating hydroponics at Ohkubo Engei

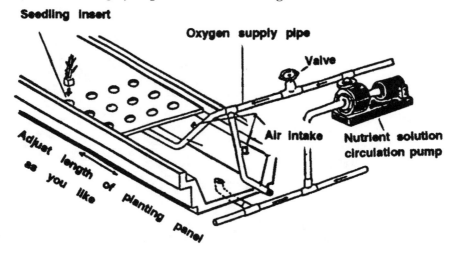

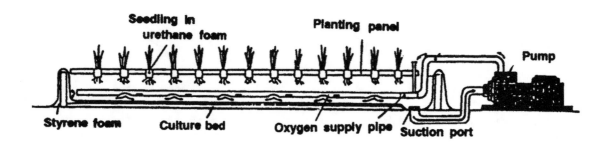

to packing stations and refrigerated storage. The upper tiers or troughs obviously shade out plants on the lower tiers, causing some reduction in yield and quality of the latter; this effect could be reduced by eliminating some of the upper tiers to permit more radiation to enter. This capital-intensive system is not yet in widespread use; its economic viability remains to be demonstrated.

A self-contained CEA module called the Ruthner System is being marketed in Austria. Two vertical conveyor belts move within a prefabricated, artificially lighted steel structure that is insulated and lined with reflective material. Plants are moved up and down within a three-dimensional light grid and supplied with nutrient solution at the bottom of the unit. As is the case with earlier CEA systems based totally on artificial light sources, it is difficult, despite marketing claims, to imagine a competitive advantage for such an energy-intensive technology in the everyday world.

Deep Flow Hydroponics. In 1976, a method for growing a number of heads of lettuce or other leafy vegetables on a floating raft of expanded plastic was developed independently by Jensen (1980) in Arizona and Massantini (1976) in Italy. Large-scale production facilities (Figure 28) are now common and are quite popular in Japan (Jensen 1989). In the Caribbean, lettuce production has been made possible by combining this system of hydroponics with cooling the nutrient solution, which stops the bolting of lettuce.

The production system consists of horizontal, rectangular-shaped tanks lined with plastic. Those developed by Jensen (1985) measured 4 m x 70 m, and 30 cm deep. The nutrient solution was monitored, replenished, recirculated, and aerated. Rectangular tanks have two distinct advantages: the nutrient pools are frictionless conveyor belts for planting and harvesting movable floats, and the plants are spread in a single horizontal plane for maximum interception of sunlight. In the Arizona trials, four commercial cultivars of lettuce were grown in the floats: a short-day leafy type ('Waldemann's Green') and three cultivars of summer butterhead ('Ostinata', 'Salina', and 'Summer Bibb'). In comparing these cultivars to similar types grown under OFA they exhibited the best tolerance to bolting, tip-burn, and bitterness, common physiological disorders which are common in warmer climates such as the tropics.

Two- to three-week-old seedlings were transplanted to holes in the 2.5 cm thick plastic (polystyrene) floats in staggered rows with approximately 30 cm/plant. (The original idea was to plant more heads with narrower spacings and to transfer them in mid-growout to floats with fewer holes and wider spacings. This concept was modeled but not executed.) Under the high light conditions of Arizona, raceway growout time to harvest was 4-6 weeks. As a crop of several floats was harvested from one end of a raceway, new floats with transplants were introduced at the other end. Long lines of floats with growing lettuce were moved easily with the touch of a finger.

Growth rates of all cultivars correlated positively with levels of available light. This correlation held to the highest levels measured, even though radiation levels in the Arizona desert are two to three times that of more temperate climates (Glenn 1984). This finding was surprising, since OFA lettuce is saturated by relatively low levels of light, and growth is inhibited as radiation increases. In addition, in other regions greenhouse lettuce is usually regarded as a cool-season crop.

A further finding was that crops grown during autumn, when daylight hours are decreasing, used available light two to three times more efficiently than winter or spring crops. Daytime air temperatures also correlated positively with growth; therefore, fall crops, grown under higher temperatures, were more efficient than spring crops. (However, during the summer monsoon season when evaporative cooling systems are ineffective, the combination of high-temperature and high light levels caused lettuce to bolt. Chilling the nutrient solution reduced bolting.) The best predictor of lettuce growth in the prototype raceway system was the product of daytime temperature and the log of radiation (Glenn 1984).

Concurrent with the operation and modification of the raceway production systems, packaging and marketing experiments were conducted. Packaging individual heads in air-sealed plastic bags extended shelf-life up to three weeks and provided protection during transportation. This procedure has also been effective in Norway (Lawson 1982). (Sealing the heads in a CO_2 atmosphere had no apparent beneficial effect.) The lettuce was also sealed with the roots intact, as researchers at General Mills (Mermelstein 1980) had reported, such packaging keeps plants alive and unwilted for extended periods. This procedure has also been used by ICI in England (Shakesshaft 1981). In the Arizona experiments, however, the roots-on package did not appear to increase shelf-life, tripled the cost of preparation and packing, increased product volume and weight for transportation, and was not particularly popular with wholesalers or retailers.

Deep flow hydroponics for lettuce production is technically sound but uneconomical in the United States, because lettuce can be grown year round in the open field at less cost per unit. Such a system may be better suited to tropical regions, where local open field production does not occur during the warmer months. In these areas, such production systems, in combination with root cooling, certainly deserve consideration.

In Norway, the interception of sunlight is maximized in the deep flow hydroponic system developed by the University of Arizona, Tucson.

Dynamic Root Floating Hydroponic System (DRF). This system of hydroponics is similar to the deep flow hydroponic system but is designed specifically for hot, tropical

Figure 29. Main components of the DRF hydroponic system

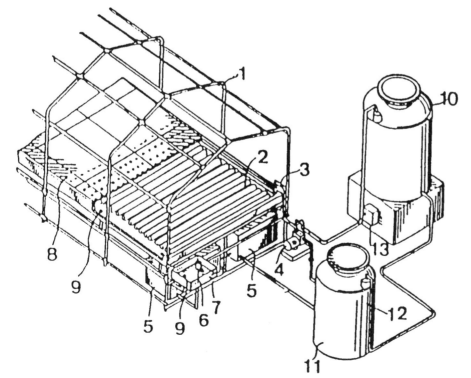

1. pipe house 2. culture bed 3. aspirator 4. pump 5. reservoir 6. nutrient level adjuster 7. nutrient exchange box 8. panel 9. nutrient outlet plug 10. upper nutrient tank 11. lower nutrient tank 12. floating switch.

regions. The major difference is the air space provided between the polystyrene growboard and the level of nutrient solution maintained in the gutter of the culture bed. Small, fluffy feeder roots, which develop in the air space, are able to absorb much of the plant's necessary oxygen requirements, as opposed to roots in the nutrient solution which is often low in dissolved oxygen. Once the temperature of the solution rises above of 25-30°C, it becomes increasingly difficult to maintain the necessary dissolved oxygen in the solution. This method of providing optimum oxygen levels to the plant may be more economical than chilling the nutrient solutions to temperatures below 25°C. (To date, there has not been a cost comparison made between the DRF system and deep flow hydroponics where the nutrient solution is cooled.)

In Taiwan, the dynamic root floating method has been coupled with a low-cost growing structure. This system of protected agriculture is recommended for growing regions that experience a high number of typhoons, heavy rainfall, high temperatures, and high insect populations (Kao 1990). The DRF hydroponic system was designed to include a gutter-shaped culture bed, an aspirator, nutrient level adjuster, nutrient exchange box and a nutrient concentration controller plus a typhoon-proof low height plastic film greenhouse (Figure 29).

The framework of the typhoon-proof low height plastic greenhouse is made of galvanized tubular iron pipe. The standard size of the plastic house is 2.13 m wide, 1.80 m high and 7.2 m in length. The length of the structure may be enlarged depending on the market demand of the produce grown. In Taiwan, the roof of the greenhouse is covered with a dew-resistant transparent PVC plastic film to prevent rain water from coming in contact with the horticultural crop. On the sides of the house, a white polyethylene plastic net prevents entry of insects. When greenhouse temperatures exceed 30°C, a black polyethylene net, providing 40% shade, is installed 10 cm above the plastic house.

Because few, if any, chemical pesticides are used, horticultural products are most often free of any chemical residue. Products also exhibit fewer problems of insects and disease. There are currently 70 DRF hydroponic installations in Taiwan covering more than 10 hectares. Interest in the DRF hydroponic system has extended to Thailand, Malaysia and Singapore (Kao 1990).

The economic feasibility of the DRF system appears promising. In Taiwan, the capital cost of the hydroponic system, excluding the greenhouse, is $116,000/ha. The pipe greenhouse itself is an additional $208,000/ha. Economic return approximates an annual net profit of over $100,000 per hectare.

Aeroponics. In an unusual application of closed system hydroponics, plants are grown in holes in panels of expanded polystyrene or other material. The plant roots are suspended in midair beneath the panel and enclosed in a spraying box (Figure 58). The box is sealed so that the roots are in darkness (to inhibit algal growth) and in saturation humidity. A misting system sprays the nutrient solution over the roots periodically. The system is normally turned on for only a few seconds every 2-3 minutes. This is sufficient to keep roots moist and the nutrient solution aerated. Systems were developed by Jensen in Arizona for lettuce, spinach, and even tomatoes, although the latter was judged not to be economically viable (Jensen and Collins 1985). In fact, there are no known large-scale commercial aeroponic operations in the United States, although several small companies market systems for home use.

The A-frame aeroponic system developed in Arizona for low, leafy crops may be feasible for commercial food production. Inside a CEA structure, these frames are oriented with the inclined slope facing east-west. The expanded plastic panels are standard size (1.2 m x 2.4 m), mounted lengthwise, and spread 1.2 m at the base to form an end view equilateral triangle. The A-frame rests atop a panel-sized watertight box, 25 cm deep, which contains the nutrient solution and misting equipment (Jensen and Collins 1985). Young transplants in small cubes of growing medium are inserted into holes in the panels, which are spaced at intervals of 18 cm on center. The roots are suspended in the enclosed air space and misted with nutrient solution as described previously.

An apparent disadvantage of such a system is uneven growth resulting from variations in light intensity on the inclined crops. An advantage of this technique for CEA lettuce or spinach production is that twice as many plants may be accommodated per unit of floor area as in other systems; i.e., as with vine crops, the

Figure 30. Aeroponic A-frame unit, developed at the University of Arizona, makes better use of greenhouse space

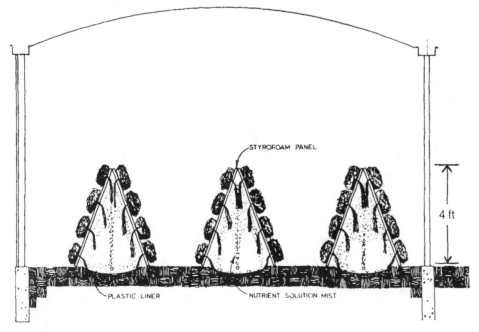

STYROFOAM PANEL

4 ft

PLASTIC LINER   NUTRIENT SOLUTION MIST

Source: (Jensen and Collins 1985).

cubic volume of the greenhouse is better utilized. Unlike the small test systems described here, larger plantings could utilize A-frames more than 30 m in length, sitting atop a simple, sloped trough that collects and drains the nutrient solution to a central sump. Greenhouses could be designed to be much lower in height (Figure 30).

Another potential commercial application of aeroponics, in addition to the production of leafy vegetables in locations with extreme space and/or weight restrictions is the rooting of foliage plant cuttings. Such a rooting system works well to control foliage diseases, and is especially important if export requirements dictate that roots of cuttings be soil-free at the time of shipping. While the cuttings require heavy shading at the time of rooting, overhead misting is not required. This greatly reduces the problems of fungal diseases and the leaching of nutrients from the foliage of the cuttings.

Figure 31. Nutrient introduction in an open hydroponic system. Plan A uses a fertilizer proportioner; Plan B uses a sump mixing tank.

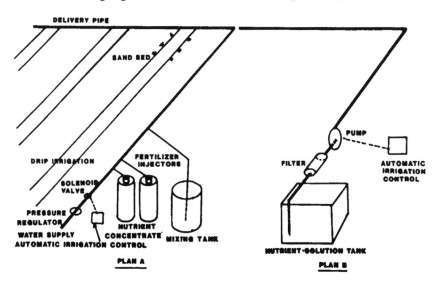

Aggregate Hydroponic Systems. In aggregate hydroponic systems, a solid, inert medium provides support for the plants. As in liquid systems, the nutrient solution is delivered directly to the plant roots. Aggregate systems may be either open or closed, depending on whether surplus amounts of the solution are to be recovered and reused. Open systems do not recycle the nutrient solution; closed systems do.

Figure 32. Open aggregate system using above ground, water-proofed troughs and drip irrigation

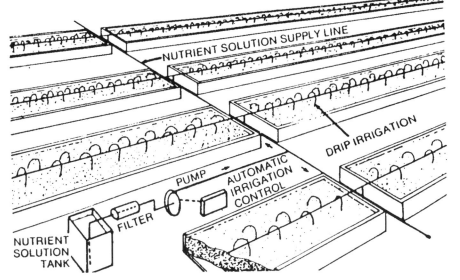

Open Systems. In most open hydroponic systems, excess nutrient solution is recovered; however, the surplus is not recycled to the plants, but is disposed of in evaporation ponds or used to irrigate adjacent landscape plantings or windbreaks. Because the nutrient solution is not recycled, such open systems are less sensitive to the composition of the medium used or to the salinity of the water. These factors have generated experiments with a wide range of growing media and the development of more cost-efficient designs for containing them. In addition to wide growing beds in which a sand medium is spread across the entire greenhouse floor, open systems may use troughs, trenches, bags, and slabs of porous horticultural grade rockwool.

Fertilizers may be fed into the proportioners (Figure 31, Plan A) or may be mixed with the irrigation water in a large tank or sump (Figure 31, Plan B). Irrigation is usually programmed through a time clock. In larger installations, solenoid valves are used to irrigate only one section of a greenhouse at a time. This permits the use of smaller sized mechanical systems.

Trough or Trench Culture. Some open aggregate systems involve relatively narrow growing beds, either as above ground troughs (Figure 32) or subgrade trenches, whichever are more economical to construct at a given site. In both cases, the beds of growing media are separated from the rest of the greenhouse floor and confined within waterproof materials. For ease of description, this system

Cucumbers, along with tomatoes, are the principle crops grown in in rockwool; a method originating in Denmark.

will be referred to as trough culture.

Concrete has been traditionally used for construction of permanent trough installations. (Sometimes it is covered with an inert paint or epoxy resin). Fiberglass, or plyboard covered with fiberglass, is also used. Polyethylene film, at least 0.01 cm in thickness, is now commonly used to reduce costs. The film, usually in double layers to avoid leakage (pinholes in either layer will seldom match up),is placed atop a sand base and supported by either planks, cables, or concrete blocks.

The size and shape of the growing bed are dictated by labor efficiencies rather than by engineering and biological constraints. Vine crops such as tomatoes usually are gown in troughs wide enough for only two rows of plants; this permits ease in pruning, training, and harvesting. Low growing plants are raised in somewhat wider beds, with a midpoint a worker can conveniently reach at arm's length. Bed depth varies with the type of growing media; about 25 cm is a typical minimum. Shallower beds of 12-15 cm are not uncommon, but these require close attention to irrigation practices. Length of the bed is limited only by the capability of the irrigation system, which must deliver uniform amounts of nutrient solution to each plant, and by the need for lateral walkways for work access. A typical bed length is about 35 m. The slope should have a drop of at least 15 cm per 35 m for good drainage; there should be a well-perforated drain pipe of agriculturally acceptable material inside the bottom of the trough, beneath the growing medium.

Because open systems are less sensitive than closed systems to the type of growing medium used, there has been much regional ingenuity in locating low-cost inert materials for trough culture. Typical media include sand, vermiculite, sawdust, perlite, peat moss, mixtures of peat and vermiculite, and sand with peat or vermiculite.

Bag Culture. Bag culture is similar to trough culture, except that the growing medium is placed in plastic bags. These bags are placed in lines on the greenhouse floor, thus avoiding the costs of troughs or trenches and of complex drainage systems.

The bags may be used for at least two years and are much easier and less costly to steam sterilize than bare soil.

The bags are typically made of UV-resistant polyethylene, have a black interior and will last in a CEA environment for two years. The exterior of the bag should be white in deserts and other regions of high light levels; this will reflect radiation and inhibit heating of the growing medium. Conversely, a darker exterior color is preferable in northern, low-light latitudes to absorb winter heat. Bags used for horizontal applications are usually 50-70 liters in capacity. Growing media for bag culture include peat, vermiculite, or a combination of both, to which may be added polystyrene beads, small waste pieces of polystyrene, or perlite to reduce the total cost. In Scotland, pure perlite is being used increasingly in bags. When placed horizontally, bags are sealed both ends after being filled with the medium.

It is beneficial to cover the entire floor with white polyethylene film before placing the bags. Jensen demonstrated with trough culture in New Jersey that 86% of the radiation falling on a white plastic floor is reflected back up to the plants compared with less than 20% of the light striking bare soil (Jensen and Collins 1985). Such a covering may also reduce relative humidity and the incidence of some fungal diseases.

Paired rows of bags are usually placed flat, 1.5 m apart from center to center with some separation between bags; i.e., each row is end-to-end. This is the normal row spacing for vegetables. Holes are made in the upper surface of each bag for the introduction of transplants, and two small slits are made low on each side for drainage or leaching. Some moisture is introduced into each bag before planting. Less commonly, the bags are placed vertically with open tops for single-plant growing. These bags are less convenient to transport, require more water, and maintain less even levels of moisture.

Drip irrigation of the nutrient mix, with a capillary tube leading from the main supply line to each plant, is recommended. Plants growing in high-light, high-temperature conditions will require up to 2 liters of nutrient solution per day. Moisture near the bottom of the bagged medium should be examined often, and it is best to err on the wet side.

The most commonly grown crops in bag culture are tomatoes and cucumbers, as well as cut flowers. When tomatoes are grown, each bag is used for two crops per year for at least 2 years. It has not yet been established how many crops may be grown before the bags must be replaced or steam-sterilized, but the sterilization can be performed by moving the bags together under a tarpaulin, at an estimated cost of less than $1,000/ha. The use of bag culture is greatly dependent on the availability and cost of growing media. To import such materials can be cost prohibitive.

Rockwool Culture. Horticultural rockwool is becoming increasingly popular as a growing medium in open hydroponic systems. Rockwool systems now receive more research attention than any other type in Europe. Cucumbers and tomatoes are the principle crops grown in rockwool; in Denmark, where rockwool culture originated in 1969, virtually all cucumber crops are grown on rockwool. This technology is the primary reason for the rapid expansion of hydroponic systems in Holland. There, the systems increase has been from 25 ha in 1978 to 80 ha in 1980, to more than 500 ha by the end of 1982 (Van Os 1983).

In the Westland region of Holland, which has the highest concentration of CEA greenhouses in the world, soil was used and steam sterilized as necessary until the late 1970's. Until then, inexpensive natural gas was available for sterilization. When fuel costs increased, Westland growers turned to methyl bromide as a soil fumigant. However, when bromides began to build up in the groundwater (and salinity increased due to saltwater intrusion from expansion of regional ship canals), so the Dutch government began to curtail the use of methyl bromide (Van Os 1983). Lacking other inexpensive means of soil sterilization, increasing numbers of Dutch operators turned to hydroponics, experimenting first with such growing media as peat and even bales of straw, with NFT and bag culture, until the rockwool culture was developed. Rockwool is manufactured by specialized companies and made from molten rock or slag (see Glossary).

As a growing medium, rockwool is relatively inexpensive, inert and biologically nondegradable. It absorbs up water easily, is approximately 96 percent "pores" or interstitial air spaces,

Figure 33. Typical layout for rockwool culture. Rockwool slabs are positioned for ease of drainage and use of bottom heating.

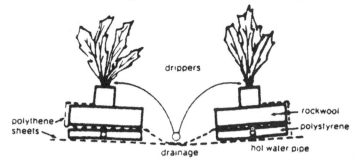

VIEW END OF BED

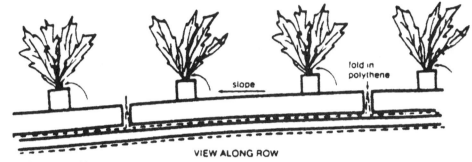

VIEW ALONG ROW

Source: (Hanger 1982b).

has evenly-sized pores (which has important consequences for water retention), lends itself to simplified and lower-cost drainage systems, and is easy to bottom-heat during winter. Its versatility is such that rockwool is used in plant propagation and potting mixes, as well as in hydroponics. A typical layout for open-system rockwool culture is shown in Figure 33.

Rockwool has several inherent advantages as an aggregate. It is lightweight when dry, easily handled, simple to bottom heat, and easier to steam-sterilize than many other types of aggregate materials. It can be incorporated as a soil amendment after crops have been grown in it for several years. In addition, an open system with rockwool permits accurate and uniform delivery of nutrient solution, requires less equipment, fabrication, and installation costs, and entails less risk of crop failure due to the breakdown of pumps and recycling equipment. The obvious disadvantage is that rockwool may be relatively costly unless manufactured within the region. Because it is nondegradable, rockwool presents disposal problems. In Holland, researchers are seeking ways to mix the used rockwool with cement for the manufacture of lightweight building blocks. The rockwool may also be reconstituted for reuse in crop production. There is a great deal of emphasis on developing a cultural system that requires less rockwool, as explained in the upcoming section entitled "NFT and Rockwool."

Plants in small rockwool blocks are set atop holes cut in plastic covering of rockwool slabs. The plant is drip irrigated with nutrient solution.

Sand Culture. Concurrent with the beginning of rockwool culture in Denmark in

Greenhouse pepper plants grow in pure sand in a facility designed by the University of Arizona on Kharq Island, Iran. Plants are drip irrigated with nutrient solution.

ingly.

Different types of desert and coastal sands with various physical and chemical properties were used successfully by the University of Arizona workers. The size distribution of sand particles is not critical, with the exception of very fine materials, such as mortar sand, which does not drain well and should be avoided. If calcareous sand is used, it is important to maintain a nutrient solution of neutral pH. Increased amounts of chelated iron must be applied to the plants. Sand growing beds should be fumigated annually because of possible introduction of soil borne diseases and nematodes.

Irrigation practices are particularly critical during the high-radiation summer months, when irrigation may be necessary up to eight times per day. Proper irrigation is indicated by a small but continuous drainage, 4-7 percent of the application, from the entire growing area. Evaporation of water around small summer tomato transplants is often high, which can lead to a slight buildup of fertilizers in the planting beds. Extra nitrogen causes excessive vegetative growth, thus decreasing the number of fruits. This can be avoided by reducing the amount of nitrogen in the solution from the time of transplant until the appearance of the first blossoms. Drainage from the beds should be tested frequently, and the beds leached when drainage salts exceed 3000 ppm.

The principle crops grown in sand culture systems are tomatoes and cucumbers, and yields of both crops have been high. Seedless cucumber production has exceed 700 MT/ha.

1969, a type of open-system aggregate hydroponics, initially for desert applications and using pure sand as the growing medium, was under development by researchers at the University of Arizona (Jensen 1973). It was logical to investigate such potential. Because other types of growing media must be imported to desert regions and may require frequent renewal, they are more expensive than sand, a commodity usually available in abundance.

The Arizona researchers designed and tested several types of sand-based hydroponic systems. The growth of tomatoes and other greenhouse crops in pure sand was compared with the growth in nine other mixtures (e.g.,sand mixed in varying ratios with vermiculite, rice hulls, redwood bark, pine bark, perlite, peat moss, etc.). There were no significant differences in yield (Jensen and Collins 1985). Unlike many other growing media, which undergo physical breakdown during use, sand is a permanent medium. It does not require replacement every 1 or 2 years.

Pure sand can be used in trough or trench culture. However, in desert locations, it is often more convenient and less expensive to cover the greenhouse floor with polyethylene film and a system of drainage pipes (PVC pipe 5 cm in diameter, cross cut one third through every 45 cm down the length of the pipe with cuts faced down) and then to backfill the area with sand to a depth of approximately 30 cm. If the sand bed is shallower, moisture conditions may not be uniform and plant roots may grow into the drain pipes. The areas to be used as planting beds may be level or slightly sloped. Supply manifolds for nutrient solution must be sited accord-

Greenhouse eggplants are grown on beach sand along the Gulf of California in a cooperative research project of the University of Arizona and the University of Sonora, Hermosillo, Sonora, Mexico.

CLOSED SYSTEMS

Gravel. Closed systems using gravel as the aggregate material were commonly employed for commercial and semi-commercial or family hydroponic CEA facilities 20 years ago. For the most part, these installations have been superceded by NFT systems or by the newer, open aggregate systems described in the previous section. As in all closed systems, a water agricultural suitability analysis must be done. Great care must be taken to avoid the buildup of toxic salts and to keep the system free of nematodes and soilborne diseases. Once certain diseases are introduced, the infested nutrient solution will contaminate the entire planting. Such systems are capital intensive because they require leak proof growing beds, as well as subgrade mechani

cal systems and nutrient storage tanks. Gravel is not recommended for use as an aggregate in a closed system.

NFT and Rockwool. A more recent development in Europe is the combination of NFT and rockwool culture, which is a closed aggregate system. In this application, plants are established on small rockwool slabs positioned in channels containing recycled nutrient solution. Compared to the open rockwool system previously described , this procedure reduces the amount of rockwool required, a great advantage when discarding the used rockwool. The rockwool, in combination with the NFT systems, acts as a nutrient reservoir in case of pump failure and helps to anchor the plants in the nutrientchannel.

The principle greenhouse crops grown in Abu Dhabi, United Arab Emirates, were tomatoes and cucumbers. The facility was developed by the University of Arizona, Tucson.

6

FLORICULTURE CROPS

As with vegetable production, the growing of floricultural crops in greenhouses is high technology and often capital-intensive. The major greenhouse floriculture crops in the United States are potted chrysanthemums, poinsettias, and Easter lilies, as well as fresh cut flowers for floral designs. Over 135 different species are grown as cut flowers; many of these grown as important export crops. Potted perennial and biennial crops, form another part of the floricultural industry. In the past decade, bedding plants have become an increasingly important floriculture product.(Tayama and Roll 1989).

Every floricultural crop has its own specific cultural and management requirements; these include stock plant management, propagation, nutrition, light intensity and photoperiod, temperature, carbon dioxide, irrigation, water quality, crop scheduling, and pest control, plus many other cultural areas.

Possibly the best of the many excellent publications on producing floricultural crops is published by Ohio State University, Columbus, Ohio.

In general, the growing media requirements for all floricultural crops is basically the same, whether it be the production of cut flowers or potted plants. For potted plants there are several different methods of irrigation. The greenhouse designs and environment control systems are basically like those systems used in greenhouse vegetable production, except growing benches are common in floriculture production and not in vegetable production.

GROWING MEDIA AND MIXING

Selection of the appropriate growing media for floricultural crops is important, not only for the grower but for the consumer who ultimately purchases the plant. The choice of a growing medium, whether mineral soil based or a soilless mixture, can mean success or failure in producing quality plants. A soil mixture must be light, porous, well-drained, of moderate nutrient content and easy to manage.

When selecting an appropriate growing medium, the advantages and disadvantages of each of the factors listed in Table 16 must be considered.

Table 16. Characteristics to consider when selecting a growing medium (Tayama and Roll 1989)

Availability	Mixing
Versatility	Pasteurization or
Physical characteristics	sterilization
Uniformity (reproducible)	Storage
Chemical residues	Potting ease
(herbicides, salts)	Growing ease
Ease of use	Fertilization requirement
Weight	Soluble salts tolerance
Equipment needed	Cost

It is easy to underestimate the cost of mixing a growing medi-

um. For example, a l:l:l mixture of mineral soil, sphagnum peat moss,and perlite may be quite expensive. But cost alone should not determine the selection: all properties of the growing medium are important to the success of production.

There is no one recipe for the ideal growing medium. The best mixture is that one which meets specific needs and is suited to the operation. Most mineral (field) soil based and soilless mixtures contain either sphagnum peat moss, vermiculite or perlite, wood by-products, or calcined clay to increase the water holding capacity and aeration, and to prevent the growing medium from becoming compacted. These various constituents have certain advantages under different circumstances (Tayama and Roll 1989).

When preparing a mineral soil-based mix, a sample of the proposed medium should be sent to a laboratory for an analysis. The analysis should indicate the pH and nutrient status and provide recommendations for any needed corrections.

According to P. Allan Hammer of Purdue University in a publication edited by Tayama and Roll (1989), the suggested proportions for a mineral soil-based mixture are:

1. In heavy, clay type field soils, a 2:3:3 or 3:5:5 (soil:sphagnum peat moss:perlite or calcined clay, by volume) works well.
2. In lighter field soils, a 1:1:1 or 1:1:0 (soil:sphagnum peat moss:perlite or calcined clay, by volume) may have better water holding characteristics.

A mineral soil based growing medium should be pasteurized before use. Steam or aerated steam is the best method of medium pasteurization or sterilization. Heat the mixture to 71°C for 30 minutes. Chloropicrin or methyl bromide-chloropicrin mixture can be used for sterilization of the growing media. These chemicals require special application equipment; the procedure must be done out-of-doors during warm (16°C) weather.

In the production of potted plants, a growing medium containing a field soil has proven beneficial. Mixes with mineral soil do not dry out as quickly and are not as top-heavy when handling as compared to soilless media.

Although some growers will purchase premixed growing media, many growers prepare their own. They can be mixed by hand, or with the two types of equipment which are available: batch and continuous flow.

Concrete and drum mixers are available in capacities of 1 to 10 cubic meters. A revolving drum or agitators supply the necessary mixing action. It is advisable to have a mixer that will enable self-unloading and steam pasteurization, and has variable speeds. Care must be taken to insure a homogeneous mix without excessive pulverization, especially that of vermiculite, or ball formation (the forming of a small ball of mix). Complete mixing should be accomplished in less than two minutes.

Shredders are often used for mixing. A disinfected, paved floor or pad area is desirable for ease in handling. Shredders have the disadvantage of producing fine particles. Roto-tillers can be used for mixing by placing alternate layers of components on a paved surface and driving the tiller slowly over the pile several times. It is generally difficult to get a uniform mix with this system, especially if small amounts of micro-elements are used.

One system that provides a continuous quantity of mix for a flat or pot filling line uses feeder bins, an open-ended drum mixer and belt conveyors. One to five cubic meter bins are available. The mixer receives the metered components from the feeder bins at one end, mixes it with a tumbling action and discharges it to a conveyor at the other end. Rates of up to 50 cubic meters per hour are possible.

Many flat and pot filling systems are available. All machines can be adapted to various container sizes, within certain limits. Filling rates vary but are generally in the range of 10 to 30 flats per minute and 20 to 50 pots per minute. Most machines require 2 to 3 people to operate them efficiently. Large growers can realize significant savings by using these mechanical fillers, since the production rate per man hour can be three to four times greater than when filling by hand.

CUT FLOWER PRODUCTION

Cut flowers may be produced in the open field or in a greenhouse. Open field production is mostly seasonal; greenhouse production can be year round for both warm and cool season floricultural crops. Greenhouse production is much more efficient and predictable than open field plantings, which are fast disappearing because of a lack of control over the growing environment. The problems of rain, disease, insects and weeds have served to make such plantings very disappointing in comparison with products produced in greenhouses.

Growing Systems. Cut flowers are grown in either ground beds or in raised 1.3 to 1.7 m wide benches. The benches may either be level or V-shaped, water tight or non-water tight. In general, the water tight bench requires the V-shaped bottom, with a drainage system installed at the bottom of the V. The drain may be PVC pipe, tile, or just gravel. With this type of bench, the crops can be irrigated by maintaining a constant level of water in the bottom of the bench, or by running water through the drain pipe or tile or through an above ground irrigation system. The non-water tight bench is either V-shaped or flat and requires a layer of sand or gravel at the bottom. Greenhouse producers in the United States are currently facing increasing pressure from regulatory agencies to insure that drainage water from the growing beds does not enter the open soil, to prevent contamination of ground water from leachate containing agricultural chemicals.

Several types of above ground **irrigation systems** are used depending on the crop requirements. These may be overhead sprinklers, or spray type nozzles, mist or fog systems. Mist or fog systems are primarily used for propagation or for cooling. Another choice could be a system of surface irrigation: spot-spitters or drip irrigation. These systems are usually automated through the use of solid state electronics, electric timers or even computers. Such irrigation equipment may be too expensive for many developing communities. Hand watering may be necessary in these areas, but care must be taken in wetting the plants. Wetting the floral portions of the plant may encourage diseases and poor quality water will spot both leaves and flowers. Wherever possible, growers should use drip lines or emitters for irrigation.

For the sides of ground beds or raised benches, several types of materials are used: wood, concrete, transite (a mixture of asbestos and concrete compressed into thin sheets), aluminum alloy, steel, tile, or fiberglass. In most countries, the wooden bench is the most common, the least durable, and in the long run, the most expensive.

The post-harvest handling and prepackaging of floral crops is most important. The fundamental processes concerned are: (1) water absorption and transpiration and (2) respiration. The principle environmental factors are: (1)temperature, (2) relative humidity, and (3) air-movement. Since cut flowers can absorb water only through the stem, the area of absorption is exceedingly small compared to the area of transpiration. Thus, the utilization of environmental factors that reduce the rate of transpiration will simultaneously reduce water deficits and prolong the life of the cut flowers. Certain formulations are used as treatments to extend the life of the flower. These consist of biocides, floral preservatives, sugar solutions, and silver thiosulfate.

POTTED PLANT PRODUCTION

The growing of plants in pots and similar containers differs from the growing of plants in the greenhouse bed or bench. The volume of soil in the container is relatively small, the root system is restricted, and the natural nutrient supply limited. The necessity for frequent watering must be skillfully managed to provide uniform water and nutrient availability for optimum plant growth and development.

Once pots for planting were made of clay, but today most are plastic. The types of pots are: (1) standard (width of the top and height are the same), (2) pan (height equals one half of the width), (3) azalea (height equals three- quarters of the width), and (4) rose (height equals one and one quarter of the width).

In general, plants growing in pots and other containers may be divided into two groups: (1) those grown primarily for their flowers and secondarily for their foliage, and (2) those grown for their foliage only.

Growing Benches. On occasion, plants will be grown on the greenhouse floor, of solid or porous concrete, in which hot water pipes are installed to give bottom heat. Heating pipes are especially important in northern latitudes during winter months. Sometimes a floor of plastic/vinyl material replaces the concrete. Porous concrete makes a very good floor surface for greenhouses because it allows water to pass through, thereby eliminating the puddles or standing water common to floors of regular concrete, sand, or gravel. Potted plants, or any type of container, should not be placed on unsterilized soil, sand, or gravel, to avoid soil-borne diseases.

Plants are usually grown on benches to facilitate easier handling and disease control. Movable, or rolling, benches are commonly used today for maximum utilization of greenhouse space (Figure 34). In addition to increasing growing space by 10 to 25 percent, movable benches reduce labor in greenhouses used for potted plants, nursery stock, and bedding plants.

Narrow folding conveyors are available to move plants to and from the growing area, as are monorail trolley conveyers, as illustrated on page 47. Both systems work well. An alternate design utilizes pallet trays 1.2 to 2.6 m wide by 2 to 2.8 m long that are handled on roller conveyors, tracks, or with a lift truck. The trays may be moved to a work area for transplanting, potting, and shipping.

Figure 34. Movable benches. Left to right: center aisle open, interior aisle open

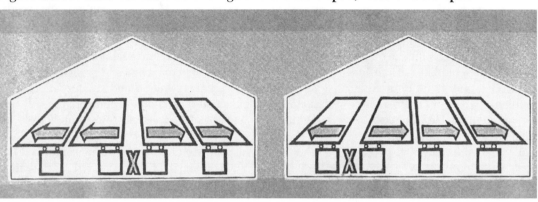

The basic concept of the movable bench system is to convert all except one aisle to growing space (Figure 34). The bench tops are supported on pipe rollers and allowed to move sideways 7 to 10 cm, the width needed for a work aisle. When there is a need to get to a particular bench, other benches in the house are pushed together, leaving the aisle at the bench. Only one side of the bench can be worked on at a time. Because the benches move, connections for water, heat, and electrical systems that are attached to the bench are made flexible. Benches as long as 65-70 meters can be moved easily by turning one of the support rollers with a crank at the end of the bench. By converting to a moveable bench system and allowing the usable space to be 80 percent of the total floor area, energy inputs per unit of product are greatly decreased in comparison to the unit cost of products from greenhouses with stationary benches.

There are many variations in bench design. The bench can be fabricated of wood or metal with either a solid or mesh bottom. Benches have been made by using concrete blocks as support legs with snow fence material (lath strips bound by wire) laid on angle iron supported by the blocks. Bench tops are also made from corrugated cement asbestos board, there where cement asbestos board, along with galvanized steel angles, are used for side boards on the sides of the bench, as well. Several manufacturers make an aluminum extrusion that adapts to an expanded metal bottom. A molded polyethylene grow tray which fits into the angled aluminum bars is also available.

Irrigation Systems. Irrigating greenhouse floriculture crops remains the most difficult task for production managers and is difficult to teach to employees. Deciding when to irrigate is still an **art** needing a "green thumb," rather than a **science**. Growers still make decisions based on sensory criteria, such as touching the medium (dry to the touch), looking at the medium (appears dry), and foliage color (bright and shiny-no need to irrigate; dull-need to irrigate) (Tayama and Roll 1989).

Unfortunately, most floriculture crops are irrigated, when the grower notices that the leaves are wilting. When this happens,

the plant has already undergone water stress, which decrease the growth rate. In the past, before the advent of growth regulators, water stress was used as a growth-regulating tool, especially with media containing field soil. With soilless media, it is possible for the plant to be under stress before it exhibits visible signs of wilting. The practice of withholding water to control plant height is not recommended because of negative side effects both to the plant and to soilless media, which may be very difficult to rewet once it is allowed to "dry out." A wetting agent may have to be added to the media or irrigation water.

The serious problem of over-irrigating has virtually been eliminated with soilless media.

According to Hackmann and Tayama (Tayama and Roll 1989) the main problems associated with the irrigation of potted plants are (1) allowing the plant to undergo stress repeatedly, even though it has not wilted, (2) root loss on susceptible varieties caused by over-irrigating or allowing the medium to dry out, and (3) excess soluble salts build-up caused by not applying enough volume of water/fertilizer to leach through the growing medium at each irrigation.

Various methods are used today to irrigate growing media. The majority of growers are using "spaghetti" tubes for irrigating and applying fertilizer. An increasing number of growers are using systems that conserve water and reduce run-off. Ebb and flow/flood, trough/nutrient film, capillary mats, and overhead/saucer systems are just some of the methods used to irrigate floricultural crops.

Drip irrigated potted foliage plants fill much of the greenhouse space, maximizing production of high quality plants.

Overhead Mist or Fog. These systems are used primarily for plant propagation, designed to keep the leaves of cuttings wet and the environment immediately around the plants humid. Such a system is not recommended for the actual irrigation of potted or container plants as these systems do not (and should not) adequately wet the growing mix of a crop getting ready for sale.

Figure 35. Trickle "spaghetti" tube irrigation system showing main (header pipe), tubes and pots

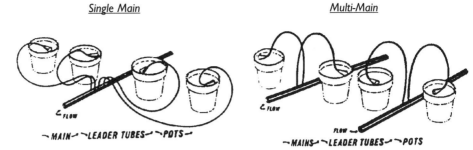

Overhead Spray and Drip. Spray nozzles on hangers, booms, and stakes wet leaves. While these systems may be suitable for irrigating crops early in their growth, it is not advisable to use such overhead watering systems later in the growth of the crop, normally after the leaves of the plant cover the entire pot. Wetting the leaves of most floricultural crops promotes the development of diseases and poor quality water will spot leaves, especially bracts of poinsettias and the floral parts of many flowering plants. Overhead watering is often used after propagation immediately after panning, the time when rooted cuttings are planted into the pot or container. Drip emitters wet leaves less and are sometimes found in hanging baskets.

Drip/Trickle Tube Irrigation. It is estimated that 80-90% of the pots produced today in the United States are irrigated by a trickle "spaghetti" tube system. Trickle tubes extending from a header-pipe to individual containers and held by weights, stakes, or other devices, do not wet leaves and are currently the preferred irrigation system used by both pot and basket producers (Figure 35).

There is a tendency to ignore day-to-day maintenance of automatic systems such as these because it is so easy to irrigate entire benches or areas at one time. Constant maintenance is required to ensure that all pots have tubes in them and that the tubes are not clogged. Good filtration of the incoming water is essential in order to remove particulates which may plug the tubes.

Subirrigation Systems. Trough, ebb and flow, nutrient film and capillary mat irrigation systems wet the growing mix by drawing water through capillary action from a source at the bottom of the container. Considerably less water and fertilizer are used with subirrigation systems than with overhead spray and trickle tube irrigation. Excess water from trough, ebb and flow, and nutrient film systems can be recycled, without water lost to the environment. This is especially important in areas with laws prohibiting runoff of excess solution containing chemical pesticides and fertilizers possible of contaminating the soil and ground water. Subirrigation systems also have the advantages of being labor-saving, and keeping the foliage dry and the water at room temperature, since the nutrient solutions are generally stored for re-use in holding tanks beneath the bench.

Since there is no leaching of subirrigation systems, excess soluble salts may build up easily in the growing medium. This can be especially true if the irrigation water has a fairly high content of extraneous salts, such as calcium, sodium, sulfur, etc., salts which are not used by plants or used in fairly low amounts.

Excess nitrogen in the foliage at harvest may give a poor post-production quality plant, one which has a poorer flower quality in comparison to those plants which are less vegetative. To counteract the build-up of soluble salts and to prevent crop losses, it is essential to use concentrated, low residue fertilizers with subirrigation systems.

Another disadvantage, as all hydroponic systems, is the potential for rapid spread of soil pathogens carried by the nutrient solution which feeds plants on the same bench. Algae may also be a problem. The initial costs, especially for the ebb and flow system, are high because of the elaborate benches and computer controls. These costs are justified, in situations where (1) advanced automation and monitoring of irrigation and fertilization are desired, and (2) it is necessary to contain runoff in order to prevent soil and ground water contamination.

BEDDING PLANT PRODUCTION

The quality of the water applied to bedding plants can greatly affect plant growth. Bedding plants suffer salt damage from water containing a large concentration of dissolved salts; therefore, it is important to know the chemical make-up of the irrigation water.

Bedding plants can be grown either on the floor of the greenhouse or on benches. For floor production, warm water pipes should be installed, in solid or porous concrete, to heat the floor. This is especially necessary in the northern latitudes, for use during winter months.

If growing benches are used, it is advisable to consider rolling/movable benches; these benches permit maximum utilization of the greenhouse space. Such benches are illustrated in Figure 34.

Irrigation Systems. Hand watering is certainly a satisfactory system, but the requirements of cost and time make it unrealistic. During periods of hot weather and rapid plant growth, it is almost impossible to satisfy irrigation needs manually.

While the hand method of water application is still common throughout the world, it has many disadvantages. It is time-consuming, requires personnel well-trained in irrigation principles and the application of water is rarely uniform. When possible, mechanized drip/trickle irrigation systems should be used.

Most bedding plants are irrigated by an overhead system. Many different, satisfactory nozzles are on the market. Most nozzles produce a relatively flat spray pattern in a 3-6 meter diameter pattern. Even the best of systems requires the use of a hose for some "touch-up" irrigation of edges and ends of benches. An overhead system should be designed for the area in which it will be used. Such equipment should be purchased from a reliable source who will provide installation information. To check the uniformity and water application rate of the system, space several containers below the sprinklers for a timed interval, and then measure the amount of water collected in each.

Probably the best and most uniform system of irrigating is with a moving boom type of equipment. This type of system is usually placed on an overhead trolley which moves back and forth over the entire area. It is propelled by an electric motor. Such irrigation systems are extremely effective but costly to install.

All irrigation systems require maintenance. Nozzles must be cleaned and replaced when they show wear. No system is completely automatic: they all require judgment in use and careful monitoring.

7

WATER SUPPLY, WATER QUALITY AND MINERAL NUTRITION

WATER SUPPLY

A correctly designed water system will supply the precise amount of irrigation water needed each day throughout the year. If a fan and pad cooling system is to be used, the amount of water required for evaporative cooling purposes must be factored into the design.

The quantity of water needed will depend on the growing area, the crop, weather conditions, the time of year and whether the heating or ventilating system is operating. Water needs are also dependent on the type of soil or soil mix and the size and type of container or bed.

In some cases it may be important to increase the estimated amount of water by 10 percent, so that leaching will reduce or prevent the accumulation of extraneous salts, salts not required by plants, such as sodium or salts not required in large amounts. Such salt build up from irrigation water is a special problem in arid regions of the world. Water should be sufficient to thoroughly wet the growing medium and to facilitate good fertilizer distribution throughout the root zone. Frequent light irrigation induces shallow rooting and may increase soluble salt concentration. Table 17 lists the estimated maximum daily water requirements for different greenhouse cropping systems.

Table 17. Estimated maximum daily water requirements (Aldrich and Bartok, 1986)

Crop	Liters of Water	Remarks
Bench crops	15 l/m^2	Based on twice daily watering (2x7.5 l/m^2)
Bedding plants	20 l/m^2	
Pot plants	20 l/m^2	
Mums, hydrangeas	60 l/m^2	Based on 3 times daily watering (3x20 l/m^2)
Roses	30 l/m^2 of bed	
Tomatoes	10 l/m^2 of bed	

A growing bed, 30 cm. in depth, with a light medium requires approximately 16 liters of water per sq. meter, per irrigation to give 10 percent leaching. Heavier growing mixes require more water. Shallower beds naturally need less water with each irrigation, but may require more frequent daily waterings.

Pot plants (15 cm. pots) need about 1/4 liter per watering. Greenhouse tomatoes, growing in a sandy loam field soil, will get a good soaking from 20-30 liters per square meter per watering. If the plants are growing in beds of pure sand, 1-2 liters per square meter may be sufficient for each irrigation but the number of waterings may reach 6 to 8 per day. Plants growing in field soil may only need 1 to 2 irrigations per week.

The maximum amount of water required per 100 square meters will vary from about 1,000 liters to 6,000 liters, for reasons mentioned earlier. The irrigation system for a greenhouse

should be able to apply the total daily needs in a 6-8 hour period. If evaporative cooling systems are used, the water requirements, added to those amounts required for irrigation, may more than double, especially in desert regions of the world with low ambient humidity conditions.

The peak use rate is the maximum flow rate of water during a 6-8 hour period. Peak use rates are needed to determine well capacity, pump capacity, pipe size, type of distribution system, and storage tank size.

If peak use rates exceed the maximum water supply or if there is a possible curtailment of the water supply, intermediate storage should be created as a back-up measure. Ponds often serve this purpose for larger growers in rural areas. However, it is crucial to make certain that the water does not contain pesticide chemicals coming into the pond from watershed runoff. Concrete or steel storage tanks should be large enough to hold at least one day's water needs. Tanks can be elevated or placed on a hill to supply water by gravity or can be connected to a pressure tank and second pump to supply water under correct pressure to the point of use. In hot climates, especially areas with strong sunlight, elevated tanks should be painted white to prevent high water temperatures. This is especially important if such storage is called upon during times of peak water use. Water temperatures over 30°C will damage plant growth if used for prolonged periods.

WATER QUALITY

Irrigation water should undergo an agricultural suitability analysis during greenhouse site selection. All water from natural sources contains some impurities: some are beneficial to plant growth; others are harmful.

Water quality has become a major concern of greenhouse growers, especially where large amounts of water are applied to a restricted volume of growing medium. Plant growth is affected by the interaction of the dissolved chemical elements in the water supply, the chemical properties of the growing medium to which the water is applied, and the fertility program employed.

Several chemical properties of water might cause problems for greenhouse growers: pH, alkalinity, soluble salts, calcium, magnesium, boron, fluoride, chloride, sulfates, sodium, carbonate, and iron.

The levels of pH and alkalinity (measured as carbonates and bicarbonates) of irrigation water affect the pH levels within the growing media. These pH levels, in turn, affect the absorption of certain nutrients by the roots and thus the health and vitality of the plant. Water pH levels above the desirable range (5.0 to 7.0) hinder absorption of some plant nutrients; pH levels below this range permit excessive absorption of certain nutrients, which may cause toxicity.

The total concentration of salts in the water is a problem common to many geographic locations. Water containing high salt levels , especially sodium, and the buildup of these soluble salts with the growing medium, results in a poor growing medium structure and reduced plant growth. The salts inhibit water uptake by the roots. While the composition of the dissolved minerals might vary according to location or source, the effect of elevated salt levels in water is the same. Boron, fluoride, chloride, sulfates, and sodium, when present in irrigation water at elevated levels, not only reduce plant quality but might influence soluble salt levels in water and growing media.

High levels of iron in irrigation water, especially in hard water containing elevated levels of calcium and magnesium, might cause brown, rusty residues that remain on leaves when overhead irrigation is practiced. This is especially detrimental when ornamental plants are grown In these situations, drip irrigation systems must be considered.

Table 18 lists the desirable ranges for specific elements in irrigation water. These general guidelines were formulated by John Peterson and Laura Kramer of the Department of Horticulture at The Ohio State University in Columbus, Ohio (Tayama and Roll 1990).

Evaporative cooling systems also require water of good quality. The alkalinity and soluble salt content of the cooling water should be determined beforehand to design the best operation and maintenance program. In evaporative cooling, the salt content of the cooling water increases in concentration as the water is evaporated. Therefore, it is important to drain a prescribed amount of water from the system to keep the salt content below the level specified by the manufacturer. The amount to drain off (referred to as "blow down rate") depends on the salt content of the water.

In Kuwait, the author has seen cooling pads completely plugged with carbonates, similar in appearance to concrete. This blockage occurred during a period of several months with the use of cooling water high in salts and without a blow-down program. The author has successfully used pure sea water for cooling with a blow-down rate of 100 percent. In other words, none of the sea water was recirculated through the cooling systems. The cooling water coming from the sea water well was returned to the bay once it passed over the cooling pads. In this case it was more economical, from an energy standpoint, to pump the water through the evaporative cooling system at a 100 percent blow-down rate than to desalt the sea water first before using it for cooling, even though desalinization would have required little or no blow-down of water from the cooling system.

MINERAL NUTRITION

Many nutrient formulas have been developed. In general, they are quite similar, differing mostly in the types of crops grown in greenhouses. It is unlikely there are any "secret ingredients" that make one formula better than another. Almost all recommendations are based on the early work of Hoagland and Arnon (1950).

There can be a significant difference in the cost, purity, and solubility of the chemicals comprising a nutrient solution, depending on the grade (pure, technical, food, or fertilizer) used. Small greenhouse operations often buy ready-mixed nutrient formulations; only water need be added to prepare the nutrient solution. Larger facilities prepare their own solutions to standard or slightly modified formulae. The commonly used weight factors in grams required to make 1000 liters of l ppm solution are given in Table 19; multiplying the factor for a chemical by the number of ppm desired in the formula will yield the number of grams to be used per kiloliter. The local availability and cost of fertilizers often determine the composition of the formula. Using Table 19, one can easily prepare virtually any formula; if a readily available chemical is not listed, its weight can be calculated from the atomic weight.

Greenhouse Vegetables. The preparation of typical nutrient solutions for tomato and cucumber culture in open or closed hydroponic systems is outlined in Table 20. A micronutrient solution designed to supplement these basic solutions is described in Table 21 (Ellis *et al.* 1974). For larger hydroponic systems, chemicals are weighed out individually to an accuracy of ±5 percent (smaller deviations generally have no apparent effect on plant growth) and arranged near the mixing tanks in a manner that precludes double-weighing of any component. The chemicals are simply added to the tanks and stirred vigorously; the order of addition is not important, but it is easiest to dissolve the most insoluble salts first (monocalcium phosphate and calcium sulfate).

Table 18. Desirable ranges for specific elements in irrigation water (Tayama and Roll 1990)

Sulfates (SO_4)	less than 240 mg/l	Soluble salts	less than l.5 mmhos
Phosphorus (P)	" " 5.0 mg/l	Zinc (Zn)	" " 5.0 mg/l
Potassium (K)	" " l0 mg/l	Sodium (Na)	" " 50 mg/l
Calcium (Ca)	" " l20 mg/l	Aluminum (Al)	" " 5.0 mg/l
Magnesium (Mg)	" " 24 mg/l	Molybdenum (Mo)	" " 0.02 mg/l
Manganese (Mn)	" " 2.0 mg/l	Chloride (Cl)	" " l40 mg/l
Iron (Fe)	" " 5.0 mg/l	Fluoride (F)	" " l.0 mg/l
Boron (B)	" " 0.8 mg/l	Nitrate (NO_3)	" " 5.0mg/l
Copper (Cu)	" " 0.2 mg/l	Ammonia (NH_4)	Undetermined
pH	5.0 to 7.0	SAR*	less than 4.0
Alkalinity	less than 200 mg/l $CaCO_3$		

*The SAR value refers to the Sodium Adsorption Ratio. This value quantifies sodium levels in relation to calcium and magnesium levels in the water. It is used to evaluate potential for growing medium permeability problems after long-term use of the irrigation water in question.

Table 19. Weight factors for calculating the amounts of chemicals (in grams) needed to prepare 1000 liters of a 1-ppm hydroponic nutrient solution

Chemical Compound[1]	Principal essential element supplied	Grams in 1000 liters
Ammonium sulfate(21-0-0)	Nitrogen	4.76
Calcium nitrate (15.5-0-0)	Nitrogen	6.45
	Calcium	4.70
Potassium nitrate (13.75-0-36.9)	Nitrogen	7.30
	Potassium	2.70[2]
Sodium nitrate (15.5-0-0)	Nitrogen	6.45
Urea (46-0-0)	Nitrogen	2.17
Nitro phoska (15-6.6-12.5)	Nitrogen	6.60
	Phosphorus	15.00
	Potassium	8.30
Monopotassium phosphate (0-22.5-28)	Potassium	3.53
	Phosphorus	4.45
Potassium sulfate (0-0-43.3	Potassium	2.50
Potassium chloride (0-0-49.8)	Potassium	2.05
Monocalcium phosphate (triple super)(0-20.8-0) 13 ca	Phosphorus	4.78
Monoammonium phosphate (11-20.8-0)	Phosphorus	4.78
Calcium sulfate (gypsum)	Calcium	4.80
Boric acid	Boron	5.64
Copper sulfate	Copper	3.91
Ferrous sulfate	Iron	5.54
Chelated iron, 9%	Iron	11.10
Manganese sulfate	Manganese	4.05
Magnesium sulfate (Epsom salts)	Magnesium	10.75
Molybdenum trioxide	Molybdenum	1.50
Sodium molybdate	Molybdenum	2.56
Zinc sulfate	Zinc	4.42

[1]Chemical compounds vary in percentage of ingredients. These figures are within the workable tolerance of most available fertilizers or chemical listed. Figures in parentheses indicate percentages of N, P, and K, respectively.
[2]2.7 g KNO_3 in 1000 liters of water equals 1 ppm K and 0.36 ppm N.

Closed Systems. Closed systems (such as nutrient film systems or NFT technique) use nutrients economically but require frequent monitoring and adjustment of the nutrient solution. Electrical conductivity is a convenient measure of the total salt concentration, but it does not indicate the concentration of major elements or the quantity of trace elements present. Thus, periodic chemical analyses are required, usually every 2-3 weeks for major elements (N,P,K,Ca,Mg,S) and every 4-6 weeks for micronutrients (Cl,B,Cu,Fe,Mn,Mo,Zn)(Graves, 1983). It is essential that the relative concentrations of nutrients in the nutrient solution approximate crop uptake ratios; otherwise, some nutrients accumulate while others are depleted. Chemical additions to the nutrient solution may be required weekly or even daily to maintain a proper balance.

Many operators, particularly of smaller closed systems, find such a schedule of monitoring, analysis, and chemical adjustment undesirable. They prefer to begin the week with a new solution; add one-half of the original formula at the end of the week, dump the remaining mixture from the tanks or sumps at the end of the second week and begin the process again. This procedure has a simplicity that may compensate for lack of precision.

Open Systems. Since the nutrient solution is not recovered and recycled in open systems, it does not require monitoring and adjustment; once mixed, it is generally used until depleted. In addition, the quality of the irrigation water is less critical than in closed systems. Up to 500 mg/kg of extraneous salts is easily tolerated; for some crops (tomatoes, for example), even higher extraneous salinities are permissible, although not desirable.

Though the nutrient solution per se does not require monitoring in open systems, the growing medium may need to be monitored. This is particularly true if the irrigation water is relatively saline or if the hydroponic facility is located in a warm, high-sunlight region. To avoid salt accumulation in the medium, enough irrigation water must be used to allow a small drainage from the planting beds. This drainage should be collected and tested periodically for total dissolved salts. If the salinity of the drainage is 3000 ppm or above, the planting beds must be leached free of salts (using the in-place irrigation system), or at least to a point equal to the salt content of the water used.

Fertilizer Proportioners (injectors). An automatically controlled open system utilizes fertilizer proportioners, manufactured devices that inject specific amounts of nutrient solution into the irrigation water. For such usage, the solution must be highly concentrated. It is prepared in two separate mixtures: one containing calcium nitrate and iron, the other containing the balance of the dissolved chemicals. Separate preparation prevents the combination of calcium nitrate and magnesium sulfate to precipitate into calcium sulfate. A twin-head proportioner is thus required. A standard design is illustrated in Figure 31, Plan A.

The rate of injection by the proportioner heads determines the necessary concentration of the nutrient solutions. For example, if each head injects one liter of stock solution into each 200 liters of water passing through the irrigation system, the stock solutions must be 200 times the concentration listed in Table 22.

Twice weekly, the total dissolved salts of the nutrient solution/water actually delivered to the plants must be checked; the proportioner pumps must be examined separately to verify that the pumps are operating correctly. Merely testing for total salts at the end of the system is not sufficient; if one head is ejecting too much solution, and the other too little, total salts may appear to be appropriate, but the ratio of elements may be badly skewed.

Nutritional Disorders. Generally, nutritional disorders relating to hydroponics are similar in cause and effect to such disorders in field agriculture. Nutritional disorders are more likely to occur in closed hydroponic systems than in open systems: impurities or unwanted ions in the recycled liquid, or from the chemicals used, may more easily destroy the balance of the formulation and accumulate to toxic levels.

Nutrient-related disorders of crop plants can be prevented by maintaining careful control of the composition of the nutrient solution, particularly in closed systems (Graves, 1983).

Table 20. Preparation of nutrient solutions for tomato and cucumber culture in closed or open hydroponic systems[1]

| | | Tomato | | Cucumber | |
| | | Soln. A | Soln. B | Soln. C2 | Soln. D |
Chemical compound (fertilizer grade)	Principal element supplied	Seedlings to first fruit set (g/1000 ltrs)	Fruit set to harvesting (g/1000 ltrs)	Seedlings to first fruit (g/1000 ltrs.)	Fruit set to harvesting (g/1000 ltrs)
Magnesium sulfate $MgSO_4.7H_2O$ (Epsom salt grade)	Mg	500	500	500	500
Monopotassium phosphate KH_2PO_4 (0-22.5-28.0)	K,P	270	270	270	270
Potassium nitrate KNO_3 (13.75-0-36.9)	K,N	200	200	200	200
Potassium sulfate[3] K_2SO_4 (O-0-43.3)	K	100	100	—	—
Calcium nitrate[4] $Ca(N)_3)_2$ (15.5-0-0)	N, Ca	500	680	680	1357
Chelated iron[5] FE330	Fe	25	25	25	25
Micronutrients[6]		150 ml	150 ml	150 ml	150 ml

1 Final nutrient concentrations in mg/kg: Soln. A-Mg(50), K(199), P(62), N(113), Ca(122), and Fe(2.5); Soln. B-Mg(50), K(199), P(62), N(144), Ca(165), and Fe(2.5); Soln. C-Mg(50), K(154), P(62), N(144), Ca(165), and Fe(2.5); Soln. D-Mg(50), K(l54), P(62), N(260), Ca(330), and Fe(2.5). All solutions are supplemented with micronutrients.

2 Solution C can be used for other vegetable crops; adjust N levels to 200 ppm for leafy vegetables as lettuce.

3 The use of potassium sulfate is optional.

4 Adjust N levels to 200 ppm for leafy vegetables such as lettuce.

5 Up to 50 g/1000 liters may be necessary if a calcareous growing medium is used.

6 See Table 21 for preparation of micronutrient solution.

Table 21. Preparation of micronutrient solution for tomato and cucumber culture in closed or open hydroponic systems[1]

Chemical compound	Element supplied	Grams to use[2]
Boric acid (H_3BO_3)	B	7.50
Manganous chloride (MnCl2.4H20)	Mn	6.75
Cupric chloride (CuCl2.2H20)	Cu	0.37
Molybdenum trioxide (MoO3)	Mo	0.15
Zinc sulfate (ZnSo4.7H20)	Zn	1.18

[1] Final nutrient concentrations in mg/kg: B(0.44), Mn(0.62), Cu(0.05), Mo(0.03), and Zn(0.09).

[2] Add water to mixture of micronutrients to make 450 ml of stock solution (heat to dissolve). Use 250 ml of this micronutrient solution with each 1000 liters of nutrient solution (Table 20)

The most common nutritional disorders in hydroponic systems are caused by too much ammonium and zinc and too little potassium and calcium. While high levels of ammonium, which cause various physiological disorders in tomatoes, can be avoided by providing no more than 10% of the necessary nitrogen from ammonium, it is best to completely avoid ammonium in nutrient solutions. Low levels of potassium (less than 100 ppm in the nutrient solution) can affect tomato acidity and reduce the percentage of high-quality fruit. Low levels of calcium induce blossom-end rot in tomatoes and tipburn on lettuce. Zinc toxicity, caused by dissolution of the element from galvanized pipework in the irrigation system, can be avoided by using plastic or other materials suitable for agriculture.

Floriculture Crops. To maximize growth, fertilization of floricultural crops during production is extremely important. Each crop requires a specific rate of fertilization. Poinsettias and chrysanthemums require relatively large amounts of fertilizer, particularly during the vegetative growth phase, while Easter lilies, which are started from bulbs, demand a different program.

Bedding plant species differ greatly in their fertilizer requirements. Those which will ultimately produce fruit, such as tomatoes and peppers, require lower nitrogen rates in order to minimize the potential problem of poor fruit set; ornamental species of bedding plants may require a higher rates of nitrogen fertilization.

With all ornamental crops, it is advisable to use fertilizer injector systems, devices which inject a small amount of fertilizer into the water line each time the plants are irrigated. Fertilizer injectors require a high level of management, which is often unavailable in many developing nations. Because injectors are subject to mechanical difficulties, they require periodic calibration and service. In some places, spare parts may be quite difficult to obtain.In these situations, it would be advisable to use bulk fertilizer solution tanks which also facilitate application of fertilizer with each irrigation.

Table 22. Amount of commonly used fertilizers applied to plants at different concentrations using a 1:100 injector. The rate is doubled for a 1:200 injector. (Tayama and Roll 1989)

Fertilizer	N	ppm P	K	Injector Concentrates **Pounds/ Gallon of Concentrate 1:100
20-10-20	200	44	166	0.8344
	250	55	208	1.0430
	300	66	249	1.2516
20-20-20	200	87	166	0.8344
	250	109	208	1.0430
	300	132	249	1.2516
15-16-17	200	93	188	1.1125
	250	109	235	1.3907
	300	141	282	1.6687
15-15-15	200	87	166	1.1125
	250	109	207	1.3907
	300	132	249	1.6687
*Potassium nitrate + ammonium nitrate	200	0	200	0.4315 + 0.3061
	250	0	250	0.5394 + 0.3826
	300	0	300	0.6471 + 0.4590
*Potassium nitrate + calcium nitrate	200	0	200	0.4315 + 0.6278
	250	0	250	0.5394 + 0.7848
	300	0	300	0.6471 + 0.9417

*A source of phosphorus should be added to these mixes. Phosphoric acid is an excellent source of phosphorus and can also be used to acidify water in those areas with high pH water. One and one-half (1 1/2) fluid ounces of 75% phosphoric acid per 100 gallons of water=44 ppm phosphorus.
**1 pound = 454.60 grams, 1 ounce = 28.35 grams, 1 gallon = 3.785 liters.

However, if the growing operation is extensive, the fertilizer solution tanks may be large and expensive to construct and will require frequent refilling. Thus, the operation of such a fertilizer system requires more labor than growing operations using fertilizer injectors. Each fertilizer application system is illustrated in Figure 31.

With injectors, approximately 200 ppm each of nitrogen and potassium are generally applied in the irrigation water for a growing medium containing field soil (Tayama and Roll 1989). For a soluble, complete fertilizer containing 20 percent nitrogen, this equals 13.4 ounces of fertilizer per gallon, or 41 pounds, 12 ounces per 50 gallons of concentrate for a 1:100 injector and twice these amounts for a 1:200 injector (1 ounce=28.35 grams, 1 pound=454 grams,1 gallon=3.785 liters). If the fertilizer contains 25 percent nitrogen, 10.7 ounces of fertilizer would be used per gallon and 33 pounds, 6 ounces per 50 gallons of concentrate for a 1:100 injector and twice these amounts for a 1:200 injector. For a soilless, or artificial, medium a higher ppm of nitrogen and potassium might be applied, depending on the floricultural crop.

The amounts of commonly used fertilizers applied to ornamental crops are shown in Table 21. Since each crop has a specific fertilizer program, including an optimum range of pH, it is advisable to follow the recommendations closely.

8

DRIP IRRIGATION

Drip irrigation, often referred to as trickle irrigation, consists of laying plastic tubes of small diameter on the surface or subsurface of the field or greenhouse beside or beneath the plants. Water is delivered to the plants at frequent intervals through small holes or emitters located along the tube.

Drip irrigation systems are commonly used in combination with protected agriculture, as an integral and essential part of the comprehensive design. When using plastic mulches, row covers, or greenhouses, drip irrigation offers the only means of applying uniform water and fertilizer to the plants. Drip irrigation provides maximum control over environmental variability; it assures optimum production with minimal uses of water, while conserving soil and fertilizer nutrients; and controls increasing water, fertilizer, labor and machinery costs.

Drip irrigation is the best means of water conservation. Generally speaking, application efficiency is 90-95 percent, compared with sprinkle at 70 percent and furrow at 60-80 percent, depending on soil type, level of field, and how water is applied to the furrows. Drip irrigation is not only recommended for protected agriculture but also for open field crop production, especially in arid and semi-arid regions of the world.

Drip irrigation is replacing surface irrigation where water is scarce or expensive, the soil is too porous or too impervious for gravity (flood or furrow) irrigation, land leveling is impossible or very costly, water quality is poor, the climate is too windy for sprinkler irrigation, and where trained irrigation labor is not available and/or is expensive.

Many types of drip irrigation systems have been developed. Row lengths may vary from a few meters on a mountainside to a thousand meters on level land. Row widths can range from one meter or less in row crops to six meters or more for orchards. The irrigation water may be relatively clean, may be drawn from open ditches with large amounts of impurities, or come from sources with high salinity. In other areas where drip irrigation is employed, water allotments may be insufficient to irrigate all the land by conventional methods. In some places, crops depend entirely upon irrigation; in others, crops require only supplemental irrigation.

Along with reduced water use, drip irrigation reduces power requirements. A typical system operates at a main line pressure of 20-30 psi or approximately 1.4-2 atmospheres. The low operating pressures permits the system to be pressurized by a deep well turbine pump. Less power means reduced energy for pumping. Since drip irrigation requires approximately half the water as compared to other irrigation methods, soluble salt concentrations will also be halved. Applying water along the plant row directs the salts away from plants to the furrows, as opposed to furrow irrigation, where salts are pushed into the root zone. Rain or a heavy irrigation will then wash the salts out of the soil profile.

Since the irrigation water is applied directly to the plant row and not to the entire field as with sprinkler, furrow, or flood irrigation, weed growth is reduced. Because the furrows, or aisles, between the plant rows remain dry, a farmer can easily enter the field with labor or tractor equipment for spraying, cultivation, or harvest.

Placing the water in the plant row

Figure 36. A typical drip system layout

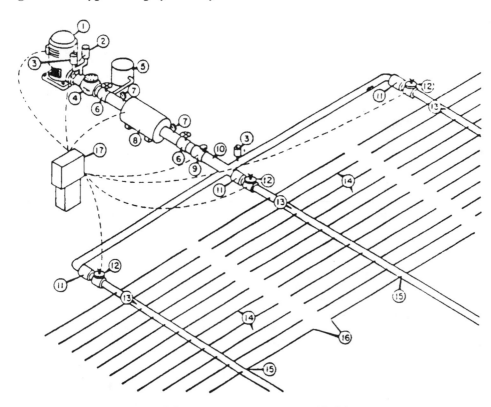

1. Pump	7. Pressure gauges	13. Submains
2. Pressure relief valve	8. Filter	14. Drip tape laterals
3. Air vents (at all high points)	9. Flow meter	15. Lateral hook up
4. Check valve	10. Mainline	16. Drain/flush valves
5. Filter injector/tank	11. Submain secondary filter (only if required)	17. System controller
6. Mainline valve/gate or butterfly	12. Field control valves manual or automatic	

increase the fertilizer efficiency since it is injected into the irrigation water and applied directly to the root zone. Plant foliage diseases may be lessened since the foliage is not wetted during irrigation.

One of the disadvantages of drip irrigation is the initial cost of equipment per acre, which may be higher than other systems of irrigation. However, these costs must be evaluated through comparison with the expense of land preparation (and continued land maintenance) often required by surface irrigation. Land leveling and canal and drain digging require heavy equipment, skilled operators, and a considerable infrastructure. Drip irrigation requires a higher level of management, not only to operate the drip system but also to maintain the fertilizer injector equipment and to keep the system properly flushed. A critical delay in the operation of a drip system may cause a decrease or a loss of crop. Frost protection that can be provided by overhead sprinkler systems cannot be achieved with a drip system. Rodents, insects, or human damage of drip tubes may cause leaks and repairs.

Maintenance of good filtration is an absolute must, since the small openings in a drip system are easily clogged. Both screen and sand filters must be checked daily and cleaned if necessary. Sand filters are easily backflushed; this operation can be automated through the use of pressure gauges located at the inlet and outlet sides of the filter. Also, precipitates that form in or on the drip lines must be eliminated: an acid flush used periodically will dissolve any mineral precipitates that may have formed in the system.

Bacteria, algae and "slime" in lines can be removed by applying chlorine or other commercial bacterial control agents applied through the fertilizer (fertilizer injector) system.

FIELD LAYOUT

Basic equipment for drip irrigation consists of a pump, a main line, delivery pipes, manifold, and drip tape laterals or emitters. The head, between the pump and the pipeline network, usually consists of control valves, couplings, filters, timeclocks, fertilizer injectors, pressure regulators, flowmeters, and gauges. Since the water passes through very small outlets in emitters, it is an absolute must that it be screened, filtered, or both, before

it is distributed in the pipe system. A typical drip system layout is illustrated in Figure 36. If the drip tape is buried, the ends of the tape should be connected to an additional submain to facilitate flushing of the lines. This is necessary because the drain flush valves are designed for surface installation of drip tape. If the system is hooked into an additional submain for flushing, it can be engineered to feed the drip tape from both submains (both ends of drip tape) during the time of irrigation. This facilitates better equalization of pressure throughout the entire length of the tape, especially with long runs.

The initial field positioning and layout of a drip system is influenced by the topography of the land and the cost of various system configurations. Design considerations should also include the relationship between the various system components and the farm equipment required to plant, cultivate, maintain, and harvest the crop.

CHEMICAL AND FERTILIZER INJECTION EQUIPMENT

Basically, there are five ways to introduce chemicals into a drip system. These are injection pumps (operated by electricity, gasoline, or diaphragm, and water-powered), venturis, pressure differential tanks, bladder tanks, and gravity. Injection pumps are normally used in greenhouse operations where a constant liquid feeding of fertilizer is required with each irrigation. The remaining four methods are not as constant and accurate in their delivery of fertilizer concentrates, but are adequate where the fertilizer or chemical is added only weekly or intermittently. Adding fertilizer or chemicals to irrigation water is often called *fertigation* or *chemigation.*

All injection pumps require energy to run. In remote areas where electricity is not available, gasoline engines and water power may be used. Piston and diaphragm injection pumps are available in various water-powered configurations (Figure 37).

A venturi injector is a tube with a constricted throat which develops a negative pressure (vacuum) at its suction port when water is passed through it at a minimum velocity. The vacuum pulls the chemical solution into the venturi port, where it is mixed with the passing water and introduced into the system (Figure 38).

Figure 37. An electrically operated piston pump

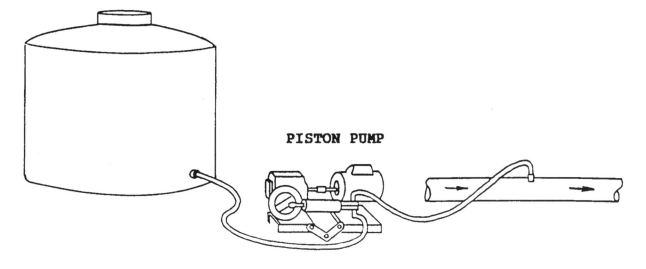

PISTON PUMP

Figure 38. A differential pressure tank and a venturi type of chemical injector

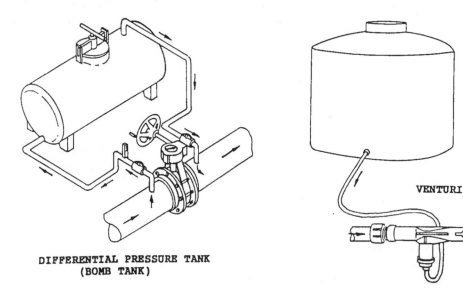

DIFFERENTIAL PRESSURE TANK
(BOMB TANK)

VENTURI

The pressure differential tank provides a simplified method to distribute fertilizer and other chemicals through a drip irrigation system. A small pressure differential created by the use of a valve or other restrictive device creates a parallel flow through the tank. The water passing through the tank dissolves and/or mixes with the material and carries it into the system (Figure 38). The only drawback is an uneven injection rate. The initial solution carried into the irrigation system is more concentrated, and is gradually diluted.

Gravity feed, when combined with a constant head device, is a very accurate way to feed chemicals into an open system. The constant head device is nothing more than a plastic box with a float valve in it. The float valve maintains a constant level of chemical which drips out the bottom through a preset metering valve. Such a system is more commonly used with furrow or flood irrigation than with drip irrigation.

A bladder tank, sometimes referred to as a proportioning tank, injects liquid materials more consistently than a pressure differential tank. The bladder tank is a pressure vessel with an inlet and an outlet opening. Inside the tank, attached to the outlet, is a bag or bladder. The bag is filled with the liquid to be injected. Water from the system is introduced into the tank, squeezing the bladder and forcing the liquid through the outlet port.

FILTERS

Media filters are tanks which contain a sandfilter bed. Sand filters are an absolute necessity with an open or surface water source. Sand filters are installed as pairs of sand-filled containers. They can be backflushed to easily clean themselves.

The need to clean or flush the filters can be determined by the loss of pressure through the filter. Therefore, pressure gauges on either side of the filter are neces-sary to indicate a loss of pressure and the need to clean or flush the filters.

If flowing water from streams or rivers is used, a sand separator filter is usually needed to remove sand from the water before it enters the sand filter.

Screen filters can be either an open or pressure type of filter. An open screen filter removes trash and other organic material from the water before it passes through the media filter. A pressure screen filter is used primarily to remove inorganic contaminants such as undissolved fertilizer salts or sand that may have escaped the sand filters. Normally water soluble fertilizer is injected into the irrigation water before it passes through the sand filter. Pressure screen filters are not suited for removal of algae or sticky organic material. Often these filters are called "polishing filters", filters that remove inorganics from the system before the water travels to the drip tapes or emitters.

PRESSURE REGULATORS

Most drip tubing is designed to operate at 0.35 to 0.68 atmospheric pressure (5-10 psi) pressure, with 0.54 atmospheres (8 psi) being the standard operating pressure. It is important to maintain even pressure to allow uniform water and fertilizer application and to avoid any chance of rupturing the drip tubes.

A spring-type (used on smaller systems) or a diaphragm-type pressure regulator can be purchased to maintain a steady water pressure in the drip system. These are inexpensive and reliable; both adjustable types or preset types are available.

In Morocco, sand filters and fertilizer injectors are set on skids to facilitate ease of movement of the system to a new location.

MAINLINES

The mainline distribution lines to the field may consist of underground plastic, pvc pipe or above ground aluminum pipe. For the submain line (header line) it is common to use vinyl "lay flat" hose. From the submain, the irrigation water flows to the drip lines, which can be either a polyethylene drip hose with inline emitters or drip tube/tape.

DRIP EMITTERS AND TAPE/TUBE

There are many types of drip emitters. In the United States there are at least eight different basic drip distribution systems, each with one or more of the following characteristics to achieve slow water discharge: tiny orifices, low pressure, or outlets with flow resistance. The eight basic systems include a flow resistance tube, a spiral resistance dripper, a porous wall hose, a small-orifice dripper, an adjustable flow dripper, a self-cleaning orifice dripper, a multiple outlet loop system and a two-pressure tube.

In Morocco, tomato crops grown on drip irrigation used 45 percent less water than sprinkler irrigated.

All are designed to reduce the outflow of water to a trickle or drip. The flow resistance tubes are small polyethylene tubes, often called "spaghetti" tubes. Connected to one end of each tube is a plastic or lead weight which anchors the tube in the container or to the base of each plant. Spaghetti tubes are commonly used for potted plants. Another common type of emitter is the spiral resistance dripper, which looks like a plastic cap over a bolt. This emitter is designed to reduce water pressure by causing the water to move along a tortuous elongated helical path. The spiral resistance dripper can be set to emit only 6 to 8 liters per hour. This type of emitter is commonly used on fruit and nut trees, on vineyard crops, and on ornamental landscape plantings.

Drip tape or tubes are used most often on row crops such as vegetables, sugarcane, and even cotton. The most common design consists of an inner and outer chamber (two-pressure tube) that distributes water evenly. The tubing is shipped flattened on a roll. Most tape is black polyethylene, 4-15 mils (100-375 microns) thick, with emitter holes at preset intervals of 20-60 cm. In general, the sandier the soil, the closer the spacing between holes. A 30 cm spacing is most commonly used in vegetable production. Most drip tapes will emit water at a ratio of 100 liters per 30 meters per hour when operated at a pressure of approximately 0.68 (10 psi) atmospheres. Only tapes with a turbulent flow emitter integrated into the drip tube are recommended. The turbulent flow design has larger passageways, which are less likely to clog and provides more uniform discharge of water on undulating terrain than tapes with conventional laminar flow designs, or designs that regulate flow through the size of the outlet orifices or by oozing through porous walls. The drip tape is most effective when buried to a depth of 20-30cm. There are drip lines in Arizona which have been buried for over ten years and are still in use. However, where soils will freeze during winter, burying is not advisable.

DRIP IRRIGATION SCHEDULES

Drip irrigation and fertilization schedules will vary with location and crop. With each irrigation, only a small amount of water is applied as needed by the plant, often on a daily or more frequent basis. This ensures a more economical use of water while obtaining maximum plant production. As the plant grows, it requires larger quantities of water because root systems become more extensive and more foliage growth is present. Therefore, water needs are increased as the growing season progresses, especially during periods of dry, hot weather.

INCREASE IN CROP YIELD

Because both water and fertilizer can be applied on a timely basis, plant stress is minimized and yields are maximized. Since drip irrigation uses less water than furrow or flood irrigation, more hectares can be watered with the same supply of water.

During irrigation trials in North Africa, it was found that drip irrigation could produce twice as many tomatoes than sprinkler irrigation using the same amount of water. In Southern California, a comparison between the effect of furrow irrigation and twin-wall drip irrigation on tomato yields indicated that drip irrigation could produce a 26.8 percent increase in total yield and a 13.7 percent increase in fruit size (Hall 1971).

Other research carried out in California has demonstrated a 12.5 percent increase in strawberry production on drip irrigated plots over furrow plots.

Drip irrigation is also becoming part of an entirely new production system for crops such as lettuce and sugarcane. For example, in the state of Hawaii, over 32,000 ha of sugar cane are now planted to drip irrigation. It has been demonstrated that sugarcane growth can be increased by 30 percent as compared to furrow irrigated fields. Undoubtedly, due to its precise control of both soil moisture and nutrient levels, drip irrigation holds great promise for orchard and row crops in the future.

DRIP IRRIGATION USED IN COMBINATION WITH PLASTIC MULCH AND ROW COVERS

Drip irrigation should be installed under plastic mulch. Early trials in New Jersey, have indicated that much higher yields of eggplant can be achieved when plastic mulch and drip irrigation are used in combination (Table 23) with each other.

Table 23. Effect of plastic mulch and trickle irrigation on eggplant yield in New Jersey

Treatment	kg/ha
Unmulched, no irrigation	66,113
Unmulched, irrigation	85,575
Plastic mulch, no irrigation	89,400
Plastic mulch, irrigation	112,912

Source: Unpub. data, James W. Patterson and N. Smith. New Jersey Agriculture Experiment Station, Rutgers University, New Brunswick, New Jersey, U.S.A.

When drip irrigation is used under plastic mulch for single row crops such as tomatoes, cucumbers, muskmelons, etc., the drip tube/tape should be placed 10-12 cm from the center of the bed and 5-8 cm deep with the emitter holes facing upward. On double row crops of summer squash, okra, eggplant, and peppers, the drip tube should be placed directly between the rows, also buried at a depth of 5-8 cm.

Many tapes are designed to prevent root growth into the emitter holes, thus eliminating this problem. Roots that do find their way into the emitters are normally removed by the flow of water during the frequent irrigations.

When laying the mulch, the soil must have sufficient moisture for seed germination. Soil to be fumigated, should be at least 10oC, be well-worked and free from undecomposed plant debris. A fumigant is used primarily for nematode control. Multipurpose fumigants offer good control of nematodes as well as soil-borne diseases. Fumigants (e.g., Vapam) can also be delivered via the drip irrigation tubing under plastic mulch. Researchers are currently exploring the use of fumigants not only to resanitize the area under the plastic mulch but also to apply herbicides to kill the first crop before a second crop is planted through the mulch (Lamont, 1991). If both the weather and soil are warm, the fumigant should escape through the plastic mulch in 12-14 days.

The combined use of drip irrigation and plastic mulch is even more important when using row covers. Furrow irrigation wets the soil under the mulch, but the capillary movement of water is not sufficient to assure good water distribution in most soils, with the possible exception of a heavy clay loam.

Experiments with tomatoes, at Cornell in 1965, demonstrated the effectiveness of the combined technologies (Table 24). Tomato crops do not always respond to a warmed soil. However, the tomato plants planted through black mulch produced greater yields with the addition of drip irrigation. Drip irrigation, when used in conjunction with all systems of protected agriculture, helps to create an environment that is ideal for maximum crop growth, while being environmentally sound, since it conserves water and inhibits leaching of chemicals.

Table 24. Early marketable production of tomatoes–kg/ha.* Cornell Uni., Ithaca, N.Y., U.S.A. (Jensen and Sheldrake 1965)

Mulch Treatment	Air Supported Row Covers	
	No drip Irrigation	Drip Irrigation
Black plastic	23,827.90	36,034.57
Clear plastic	18,130.85	26,916.54
No mulch	20,441.70	19,954.67

*12,100 plants/hectare - from four harvests with first harvest on July 22 and last August 6.

9

DISEASE AND INSECT CONTROL

It is commonly assumed that protected agriculture systems are relatively free of plant diseases and insect pests because the technology is mostly enclosed. Unfortunately, this is not true. Pathogens and insect pests may be introduced when greenhouse doors are opened and through the movement of people and materials. Indeed, pest populations can increase with alarming speed in some protected agriculture installations, such as greenhouses, because of the lack of natural environmental checks. Conversely, the enclosure of the growing area makes it easier than in open field agriculture to control disease.

For the past 50 years, crop diseases and insects have largely been controlled by chemicals. This is especially true in Europe, and in most other regions where protected agriculture is widely practiced, for both greenhouse and field crops. In these areas, many apparently effective pesticides and chemicals (none produced specifically or exclusively for protected agriculture) are available and legal.

However, in the United States, where so many of the world's agricultural chemicals have been invented, few chemicals are legal for use in greenhouses. The effects of chemicals inside CEA structures may be different and more dangerous than they are in open-field crops, and their safety must be documented before federal and state governments will certify their use. However, because of the limited use of CEA in American food production — with such a small potential market — manufacturers are unwilling to spend the large sums necessary to obtain such documentation and certification.

The frightening ability of some insects to develop resistance to chemical pesticides has revived worldwide interest in the concept of biological control: the deliberate introduction of natural enemies of insect pests, particularly when used in association with horticultural practices, plant genetics, and other control mechanisms. This combined approach, called integrated pest management (IPM), is of particular interest to Americans in CEA because of the paucity of pesticides with legal clearance for use in greenhouses.

It is absolutely imperative that growers using any system of protected agriculture have a thorough knowledge of those diseases and insects specific to the crop grown. Successful crop production requires both the identification of possible crop diseases and insect problems, and the ability to properly integrate disease and insect prevention and control practices into a total management plan.

ROOT DISEASES

Most horticultural crops are susceptible to root diseases, some more than others. Common root fungal diseases are *Fusarium* and *Verticillium* spp, *Pythium* and *Phytophthora* spp., and *Rhizoctonia* spp. A number of other root diseases also exist. Some fungi are specific to certain crops. Some crops have cultivars resistant to specific root diseases; for instance, the toma-

to is resistant to *Fusarium* and *Verticillium* spp. Other root diseases are caused by bacteria and nematodes. A medium free of soil-borne pathogens is essential to successful greenhouse production.

Closed hydroponic systems permit a rapid spread of root diseases. With tomato, the most common crop, some root death occurs naturally when the extension growth of the roots ceases as the first fruits begin to swell. At that time of natural stress, the plant is vulnerable to root pathogens (Evans, 1979) which may be introduced by contaminated seed, infected propagating mixtures, or even adjacent greenhouse soil.

FOLIAGE DISEASES AND INSECTS

There are no foliage-damaging plant diseases or insect pests associated exclusively with any system of protected agriculture. Some diseases and insects infect any one crop whether grown in the open field or under row covers or within a greenhouse. Although many chemicals have been approved in the United States for use on food crops in open field agriculture, few are registered for greenhouse applications on food crops. Many more chemicals are available for use on greenhouse floricultural crops than for greenhouse food crops.

Most disease and insect species are common throughout the world. Although some are specific to certain regions, in general, the problems and challenges are similar. In addition to those root diseases already mentioned, common foliage diseases include viruses and several other bacterial and fungal diseases.

Insects, common the world over, are aphids (family *Aphidae*), whitefly (*Trialeurodes vaporariorum, Bemisia tabaci*), fungus gnats (family, *Mycetophilidae*), shore flies (family, *Ephydridae*), thrips (*Frankliniella occidentalis, Echinothrips americana*), mealybugs, caterpillars (*Lepidoptera*), plus others.

A protected environment is ideal for the rapid proliferation of unwanted pathogens and insects. Greenhouse operations are normally free of pests for several crops after initial construction, but nearly always become infested after a period of operation.

INTEGRATED PEST MANAGEMENT (IPM)

There is no standard definition or technique of IPM. It generally consists of a carefully structured and monitored combination of biological controls, plant genetics, cultural practices, and use of chemicals. IPM guidelines are that insects can be controlled largely by encouraging their natural predators and parasites, by using resistant cultivars, and by using proper plant spacings and other cultural practices, and by using pesticides only as a last resort. Pesticides are avoided, if at all possible, because they are expensive, often reduce maximum possible yield of food crops (Burgess, 1974), and may lead to the evolution of pesticide-resistant insect species. Integrated pest management is also environmentally sound since it minimizes the dispersion of toxic materials and is consistent with the current interest in the consumption of "natural" foods. The individual components of IPM are discussed in the remainder of this chapter.

Biological Control. The use of biological control with greenhouse crops was first reported in England in the 1920s. Whitefly infestation of tomatoes was countered by the introduction of the parasite *Encarsia formosa* Gahan; results were mixed, and the practice was discontinued after the development of synthetic organic pesticides after World War II (Gould *et al.* 1975).

By the 1960s, large populations of two-spotted spider mites (a severe cucumber pest) became resistant to many pesticides. Whiteflies also began to show the same characteristic (Wardlow *et al.* 1972). The predatory mite Phytoseiulus persimilis was used successfully to control spider mites (French *et al.* 1976; Gould *et al.* 1975), and

The predatory mite is commonly used in greenhouse crops in the control of spider mites.

Encarsia once again was used against whiteflies (Gould *et al.*1975). Several other natural predators and parasites now control pests in CEA (Bravenboer 1974). Most current procedures were developed by researchers at the Glasshouse Crops Research Institute in England and the Research Station for Vegetables and Fruit Under Glass in The Netherlands. Scientists in this field meet every three years and publish a volume of conference proceedings.

An obvious difficulty with IPM is the integration of biological controls with even limited pesticide use; the chemicals usually kill the introduced enemies as well as the target pests, requiring reintroductions. In addition, it is improbable that growers could keep at hand, or have access to, predators and parasites for all major pests, or be trained to use them. In part, biological control resembles hydroponics itself: it is an attractive concept, easier to promote than to practice.

In floriculture production, it is difficult to produce foliage free of insect damage while maintaining low populations of pests, predators and parasites. Therefore, biological control is not as common in floriculture as in vegetable production.

Cultural Control. Very close spacing of plants inhibits air circulation and light penetration; this fosters the growth of insect populations and restricts access to them. Wider spacing facilitates monitoring and control of insects, as does a general program of greenhouse sanitation: removing trash, dead plant material and other debris that may shelter pests.

Physical Control. A CEA lighting system should be designed and placed to avoid attracting insects into the greenhouses. Certain interior lights such as UV or yellow mercury vapor, act as insect traps.Objects painted certain colors and sprayed with adhesive materials serve as traps for some pests. For example, boards measuring approximately 30 cm^2 and painted yellow (Rustoleum No. 659) attracted whiteflies in Arizona.

Genetic Control. For protected agriculture crops, as well as OFA crops, cultivars have been developed for tolerance or resistance to certain insects, viruses, and fungi. Disease tolerance to *Fusarium* and *Verticillium* in tomatoes has been a big boost to productivity. However, there are instances where greenhouse cucumbers, tolerant or resistant to powdery mildew (*Sphaerotheca fuliginea*), have failed to produce maximum yields. For the grower, a trade-off may be necessary between yield potential and resistance; selecting cultivars should be done case by case, taking into account region, local pest populations, etc.

Chemical Control. Chemical controls are used as a measure of last resort in an IPM program. This is more easily done in OFA, with so many approved pesticides and chemical procedures, but difficult in CEA since so few materials are certified for greenhouse use. Timeliness and uniformity of application of chemicals in CEA operations are crucial to the success of IPM programs. Greatly reduced amounts of pesticides are possible with judicious use.

10

PROPAGATION AND CULTIVAR SELECTION

With higher plants, there are two modes of propagation: (1) sexual (propagation by seed), and (2) asexual (vegetative propagation). In sexual propagation, specialized reproductive cells, called gametes, are formed in the flowers of plants. Fusion of the male and female gametes leads to the development of an embryo and eventually the seed. In asexual reproduction, new plants arise from specialized vegetative organs such as tubers, rhizomes, runners, bulbs, corms, or by various means of propagation such as rooting of plant cuttings, grafting, or layering.

Most vegetable crops reproduce by sexual means. Floricultural and nursery crops, depending on species, are reproduced by either sexual or asexual reproduction.

When selecting any food or floricultural crop, it is important to properly match the cultivar (variety) to the growing environment. The environment includes the temperature, moisture, nutrient supply, light (duration and quality), carbon dioxide, soil reaction, oxygen, wind, and pests.

Of all the factors, temperature is possibly the most important in cultivar selection. Cultivars respond very differently in different climates.

Light is also crucial to cultivar selection. While most plants do not respond to seasonal variation in light, the growth and flowering of some cultivars are affected by the length of day. This response to daylength is called photoperiodism.

Plants can be classified into three groups with respect to photoperiodism:
1. *Indeterminate* - plants whose time of flowering is not greatly affected by duration of light.
2. *Short-day* - plants that flower during short days (long nights).
3. *Long-day* - plants that flower during long days (short nights).

In general, the critical day length for long-day plants is 12-14 hours. The day length for most short-day plants ranges from 8-10 hours. Under natural conditions, the length of day varies with the month of the year, and with latitude, except at the equator.

Before introducing any cultivar into a new region, it should be tested by growing it on a trial basis, first on the plant breeder's experimental area, and then in a series of field trials. In this way, the grower can ascertain the cultivar's adaptability to the region's environmental factors, such as temperature, light, soil, etc. He should compare the new and old cultivars side by side.

Since yield and crop quality are so important in determining a grower's income, the grower must be always on the alert to use new and better cultivars. Although much of the testing can be performed by the plant breeder and the company developing the new cultivar, the grower must make trials on his own farm.

VEGETABLE CROPS

Good seed is essential to success in vegetable production, whether in protected or open field agriculture. Since CEA costs can be high it is imperative to have seed which possess-es the genetic characters suited to the environment in which it is grown.

Seed Selection. Good vegetable seed must be true to name, viable, disease and pest free, free from weed seeds, dirt and other foreign matter, and be fairly priced from the standpoint of both seller and buyer. The buyer of seed should insist on a statement of the percentage of germination; he should remember also that many kinds of seed lose viability in a relatively short time when held under unfavorable conditions.

Cultivar Selection. Cultivars suitable for open field production are usually suitable also for production with mulches and row covers. Some crop cultivars may be more tolerant of cold temperature; these would be preferred for early spring planting. In northern latitudes, relatively rapid maturing cultivars should be selected.

In Europe, tomato and cucumber cultivars are specifically bred for greenhouse production. These cucumber cultivars are unique in that they usually are only female flowering, with dark green parthenocarpic (seed free) fruit which are free of bitterness. Because the skin of the cucumber is smooth and succulent, it is easily scarred when rubbed against the rough stem of the cucumber plant. Therefore, if the fruit is to remain unblemished, it must be grown in an environment without high wind.

In the United States, some tomato cultivars are grown both in greenhouses and the open field. Tomato cultivars for greenhouse use are indeterminate, since the plant is grown upright as a single stem rather than a bush.

Certain lettuce cultivars are preferred for greenhouse production since they can tolerate the higher temperatures which sometimes occur within a protected structure with inadequate ventilation, especially during days of warm outdoor temperatures. Where high temperatures are likely, it is important to choose cultivars with good resistance to bolting (going to seed), tipburn, and plants that are bitterfree. These are all physiological disorders which are encouraged by high temperature.

Crops such as pepper and eggplant are grown interchangeably between the open field and in combination with systems of protected agriculture.

Cultivars that are tolerant to high temperature/high temperature-humidity environments, common to the tropics or deserts, are often quite different from those commonly grown in temperate regions. The University of Arizona has done extensive trials on seeds from various companies to determine the greenhouse cultivars most suitable for production in the tropics (Jensen 1972). The results of these tests are found in Table 25. The listing of seed companies does not imply endorsement. More recent cultivars than those listed, or from other seed companies, may have similar or better cultivars for a specific region. Cultivars from other seed companies, especially those conducting programs in tropical regions, and the companies listed in Table 25, should be compared.

Table 25. Cultivars suited for greenhouse vegetable production in the tropics

VEGETABLE	CULTIVAR	SOURCE
Cantaloupe	Perlita	Petoseed Co., Inc.
Chinese Cabbage	Saladeer	Takii & Co., Ltd.
	W-R Super 80	
Cucumber	Corona	De Ruiter Seeds
(Long Dutch Type)	Toska 70	Nunhems
Eggplant	Black Magic	Harris Moran
		Seed Co.
Head Lettuce	Minetto	Burpee
Bibb Lettuce	Ostinata	De Ruiter Seeds
	Salina	Len de Mas Zaden
	Summer Bibb	Harris Moran
		Seed Co.
Leaf Lettuce	Waldman's Green	Harris Moran
	Grand Rapids	Seed Co.
Pepper (Bell Type)	Takii's Ace	Takii Seed Co., Ltd.
	New Ace	
Tomato	N-65	Takii Seed Co., Ltd.

Whatever the system of protected agriculture, it is advisable to occasionally contact commercial seed companies to learn about newly-developed cultivars which might be worthy of trials. It is important to carefully describe the growing system and all factors of the environment. Seed companies are usually quite willing to cooperate in such trials, providing they receive the results.

FLORICULTURE AND NURSERY CROPS

Floriculture and nursery types of horticultural crops encompass nearly all methods of reproduction. When purchasing vegetative reproductive plant parts, it is important to buy from a reliable sales organization.

In cases where the plants from seed do not resemble the parent plants, vegetative or asexual propagation is used. For example, if seeds from a "Delicious" apple are planted, the trees which would develop will bear apples quite unlike those of the parent. They would vary greatly in size, shape, color, quality, season of maturity, and many other characteristics. But, if a bud from the tree is grafted on a piece of apple root, the tree which would grow from this bud would be identical to the tree from which the bud came. The same situation exists with many other fruit crops, and with many flowering, ornamental plants, and with certain vegetable crops as well.

It is more economical to propagate certain plants vegetatively than by producing plants from seed.

Propagation of floriculture and most nursery crops occurs in greenhouses. In warmer climates, nursery crops may be propagated in open air nurseries or within shade structures; in many cold climates, apples and pears are propagated through grafting during winter in the open field.

Because propagation techniques vary greatly, each group of

crops will be discussed briefly as they pertain to greenhouse production.

Bedding Plants. Bedding plants are mainly produced from seed. Because the cost of seed is less than five percent of the total cost of growing bedding plants, only the very best seed should be purchased. Inexpensive seed may result in poor germination, causing delays in production and lack of marketable crop during periods of peak demand.

Rapid germination of most bedding plants occurs when soil temperatures of 27°C are maintained, provided soil moisture is adequate. At temperatures of 15°C or less, germination is slow and damp-off diseases can be troublesome. There are bedding plants species which germinate best at lower temperatures. In any case, the need for exact temperature control underscores the reason for producing bedding plants within a controlled environment greenhouse.

The selection of bedding plant cultivars specific to a region is not as critical as in vegetable crop production. What is critical to the region are the dates the seed is sown. Bedding plants for outdoor plantings in the southern regions of the United States would be seeded sooner than those for use in the northern latitudes.

As a guide to planting dates, the United States is divided into 11 plant hardiness zones, according to weather conditions. Each bedding plant cultivar is planted according to dates specified in the planting zone. Countries not divided into planting zones could match their climatic conditions to similar zones in the United States and plant accordingly.

Foliage Plants. Most foliage plants originate in low light environments, under a canopy of trees and vines. They grow in warm temperatures, at varying levels of humidity. Many of these plants are easily propagated vegetatively, and make ideal houseplants. There are a number of manufactured media to stimulate the rooting.

The foliage plant business is a major greenhouse industry in the United States and Europe. Rooted cuttings of foliage plants frequently enter the United States from tropical regions in the Caribbean and Latin America, where they are propagated within shade houses. There are strict regulations against the importation of soil or plants infested with pests. With proper care, these problems can be prevented.

Flowering Potted Plants. As with foliage plants, the propagation and growing of potted plants is a major industry in the United States and Northern Europe. Most plants are propagated from vegetative cuttings; others, like Easter lilies, are from bulbs. In nearly all regions of the world, potted plant production occurs in greenhouses.

Flower growing technology is changing rapidly through total control of the greenhouse environment. Computers, robotic transplanters, chemigation, and biological controls all to help insure better quality at lower cost of production.

Plant propagation requires very specialized attention to the temperature, light levels, growing media, nutrition, and methods of irrigation. The growing technology includes the use of rooting hormones, fungicide drenches, and growth regulators, as well as special equipment to provide bottom heat to the propagation beds, and lighting and shading systems to control photoperiod along with misting systems.

Many companies specialize in the production of propagated plant material. Each company has well-defined programs for growing potted plants and regional representatives to assist growers.

Potted Perennials and Biennials. Most species of potted perennials and biennials are propagated by seed, although a few are propagated by vegetative cuttings and by bulbs.

Nursery Crops. Horticultural nursery crops can be divided into two groups: (1) young fruit and nut trees, grapes, and small fruits, and (2) young ornamental plants. The former are grown to establish fruit and nut orchards and small fruit plantations; the latter are grown to landscape or beautify homes, public and private buildings, schools, highways, parks, and industrial areas.

Most species of crops are propagated within greenhouses. In warmer climates, lath or screenhouses are used instead of greenhouses. Most species are propagated as vegetative cuttings although a few are grown from seed. According to the species of plant, the cuttings can be classified as herbaceous cuttings, deciduous softwood, and hardwood cuttings, evergreen cuttings and root cuttings. The program of propagation can involve both budding and grafting of plants. A desired fruit cultivar might be budded or grafted onto a root stock resistant to a specific disease. Some fruit species have cultivars that can be grafted onto dwarf root stock to produce dwarf trees.

PART FOUR

ECONOMIC CONSIDERATIONS

CHAPTERS II - I4

11

ECONOMICS OF PROTECTED AGRICULTURE

Regardless of the type of system, protected agriculture can be extremely expensive. These equipment and production costs may be more than compensated by the significantly higher productivity of protected agricultural systems as compared with open field agriculture.

The costs and returns of protected agriculture vary greatly, depending on the system used, the location and the crop grown. For instance, on a per hectare basis, the costs for mulching and row cover systems are insignificant when compared to greenhouse production. Furthermore, published greenhouse costs are often deceptive. Figures may include only structural and equipment costs, without adequate allowances for the expense of service buildings and offices, packing equipment and storage, construction labor, utilities, roads, fences, lighting, tools, vehicles, working capital, etc. A 1993 estimate of such a total turnkey cost in Arizona for a modern and sophisticated CEA system, exclusive of land and the interior growing system, is $85/m². If costs of a hydroponic system are included, the turnkey expense is between $90-100/m².

A Global Review of Greenhouse Food

In Norway, lettuce crops are planted through white plastic film to maximize reflection of sunlight on the crop.

Production (Dalrymple 1973) contains an excellent chapter which outlines the economic factors to be considered in planning for greenhouse agriculture. These factors remain valid; however, actual costs - especially for labor - for all systems of protected agriculture, have escalated over the last 20-30 years.

Following Dalrymple, this chapter will give an overview of issues to be considered in assessing the economic viability of a given system of protected agriculture.

By design, all protected agriculture systems of cropping are intensive in use of land, labor, and capital. Greenhouse agriculture is the most intensive system of all.

LAND USE FACTORS

The intensity of land use is greatly dependent upon the system of protected agriculture. Intensity, for economists, indicates the labor and capital requirements per unit of land. Year-round greenhouse crop production is therefore much more intensive than seasonal use of mulches and row covers. Coinciding with intensity are yields, which are normally far greater per acre from year-round than from seasonal systems.

Multiple Cropping. Multiple cropping indicates more than one crop grown per unit area of land in a single year.

Mulches and Row Covers. Multiple cropping is more common in southern than northern latitudes. In Southern California, spring tomato crops may be followed by cucumbers, planted through the same plastic mulch, with drip irrigation installed beneath the mulch. In the desert regions of the United States, melons are often planted in trenches covered with clear plastic in early January, followed by lettuce production in late summer or early fall. Prior to seeding the lettuce, the plastic cover used initially for the melons is removed. The drip irrigation system is permanent. In China, it is common to plant watermelon crops through plastic mulch after winter wheat. The wheat stubble is left for the melon vines to grow over; fewer fungal problems are encountered if the watermelon fruit rests on the wheat stubble.

In Norway, early spring lettuce crops are commonly planted through black plastic mulch for weed control, and a second crop is planted through the same mulch later in the summer. For the first crop, a floating row cover may be used in combination with the mulch in order to create a warmed micro-climate around the lettuce plant, stimulating plant growth for early harvest in summer.

Suggestions on double-cropping of vegetables using plastic mulch are listed in Table 24.

Greenhouses. Several methods of multicropping have been developed to take maximum advantage of light conditions and to respond to challenges from open field agriculture. In northern latitudes, double or triple cropping is common; in southern regions, far more crops may be grown annually, depending on the crops selected. In Arizona, where winter light levels are high, eight to ten greenhouse lettuce crops have been produced in one year. In some selected situations, ornamental crops are included in the production cycle. Bedding plants may be grown during spring, and tomatoes produced in the fall. For vegetable crop production, several different multiple cropping systems are used in the United States. Eight major patterns of greenhouse vegetable operations are outlined in Table 26 for the northern regions of America. The seasonal breakdown is only approximate, especially for the spring season, where there may be early, midseason, and late crops.

Figure 39. Average annual potential photosynthesis - grams per square meter per

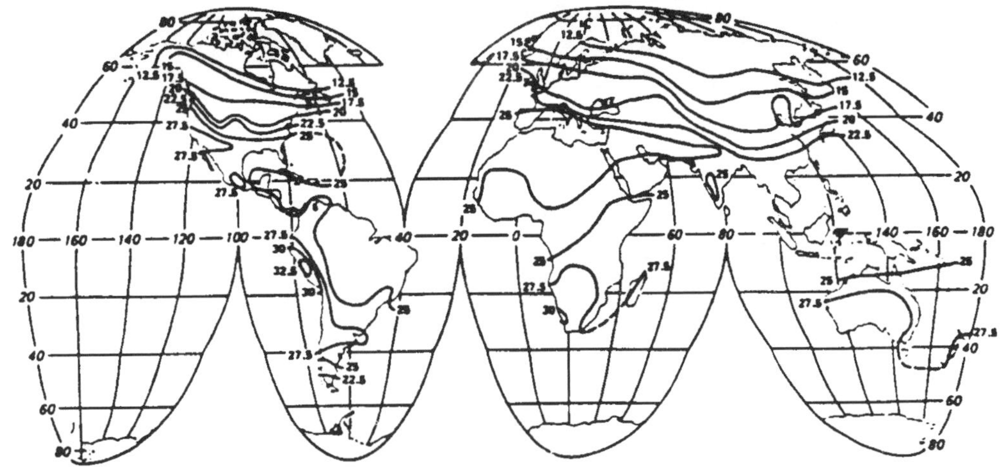

Source: (Change 1970)

Tomatoes and cucumbers are often produced through a system of monoculture. A tomato crop may be planted in February or March and repeated until late fall. Such monoculture systems may entail a substantially higher disease risk. With other crops, the rotation patterns may be somewhat mixed.

Table 26. Greenhouse vegetable rotation patterns for northern latitudes (Dalyrymple 1973)

Pattern	Fall	Winter	Spring
1			Tomatoes
2	Tomatoes		Tomatoes
3	Tomatoes		Cucumbers
4	Tomatoes	Lettuce	Cucumbers
5	Lettuce		Tomatoes
6	Lettuce	Lettuce	Tomatoes
7	Lettuce	Lettuce	Cucumbers
8	Lettuce	Lettuce	Lettuce

Spring crops of tomatoes and cucumbers normally yield more than fall crops because of more favorable light conditions in the spring. While fall crops may be unprofitable, they are grown to keep labor employed. The introduction of new European seedless cucumber cultivars into the United States has made the monoculture of cucumber crops increasingly common because they are cost competitive in the market. Standard seeded cucumbers are usually grown in the open field. When grown in a greenhouse, they do not bring the market price required to offset the higher cost of greenhouse production in competition with open field cucumbers imported from southern latitudes. Normally, two cucumber crops are grown annually, one in the fall and the second during spring/early summer; in southern latitudes three crops are usually produced annually.

Lettuce is the only crop that does well during midwinter in the northern regions. It tolerates both lower light and lower temperature conditions, although maturity time is increased. In Holland, a winter crop of lettuce may take 90 days to mature; under the higher light intensities of Arizona, a winter lettuce matures in 40-45 days. In the United States, greenhouse lettuce competes directly with open field production from Arizona, California and Mexico. Only greenhouse lettuce crops grown in the Eastern part of the United States, that produce for the surrounding market, can marginally compete with the normally excellent open field lettuce shipped in from long distance. European greenhouse growers do not face such similar open field competition during winter.

Rotation patterns in southern latitudes, where light levels (Figure 39) are far higher than in the northern latitudes, especially during winter, may be somewhat different.

Rotation patterns commonly used in desert regions, or in southern regions of the world, are shown in Table 27.

Table 27. Greenhouse vegetable rotation patterns for southern latitudes

Pattern	Mid-summer/fall (July - Oct.)	Winter (Nov.-Mar.)	Spring/early summer (April - June)
1	— Cucumbers — /	— Cucumbers – /	— Cucumbers —
2	/	Tomatoes	
3	Tomatoes	/	Tomatoes —
4	— Cucumbers — /	— Cucumbers – /	— Tomatoes —
5		Lettuce (8 - 9 crops)	

In high light regions of the world, winter production may include tomatoes and cucumbers and lettuce. In southern Europe, peppers and other vegetables may also be grown. In Arizona, tomatoes are planted into the greenhouse in late summer and allowed to produce until June or July of the following year. The packaging and quality are superior, competing well with products of less quality, grown in the open fields of Mexico and Florida.

In southern latitudes, very high yields of all vegetable crops are achieved during winter and may bring high market prices, especially in markets located in densely populated areas of northern United States.

Since the introduction of rockwool, crop rotation, especially of lettuce with cucumbers and/or tomatoes, is less common. This is due to the fact that many of the greenhouse lettuce sys-

Intercropping lettuce with tomato crops will help maximize the production of a greenhouse facility.

tems are liquid media systems, designed specifically for this crop. Growing in open soil beds facilitates the rotation of crops but these methods of production are becoming less popular, especially in many industrialized countries where labor costs are high. To compensate for high labor costs, growers are using more automation, computers, and new systems of climate control.

In the United States, the intercropping of lettuce and tomatoes was quite common years ago in greenhouse production. In those cases where three or four crops of lettuce were grown in soil, the last crop was often intercropped with tomatoes in late winter and early spring (Dalrymple 1973). With the expansion of open field lettuce production in the western part of America, such companion systems of production have become rare in greenhouse agriculture. However, intercropping lettuce with tomatoes represents a compromise as to optimal temperature for each crop.

Intercropping or companion cropping is more common in Asia than in Europe or the United States. In China, open field vegetable crops, such as tomatoes, peppers, and eggplants, are often planted through plastic mulch and intercropped with trellised cucurbit crops. Later in the summer, the vine crop shades the vegetables and protects the fruit from sunburn. However, even in Asia, intercropping is becoming less frequent because it is labor intensive. With increased family planning in mainland China, many farm communities are experiencing a decline in available labor.

YIELDS

Higher yields are normal benefits of protected agriculture over unprotected open field production. Just how much higher the yields depends on the system used and the region of production.

Mulches and Row Covers. Mulches and row covers may substantially increase early yields, although total yields may not necessarily be greater. Plastic mulches often more than double early yields; and, if row covers are used in combination with

mulch, the yield may double again.

In Chapter 3, tables 4 and 5 detail the yield differences between clear and black polyethylene mulch when cucumbers and melons are grown under row covers. Tables 9 and 10 show the yield results according to the type of row cover.

In many cases, early yields bring greater revenue. This advantage must be weighed against the cost of using plastic mulch and row covers.

Greenhouses. Balanced against the high capital and operational costs of greenhouses is the significantly higher productivity of such systems in comparison with OFA. Yield data have been reported in the literature for years; typical yields for crops grown hydoponically in desert greenhouses in the American Southwest and in the Middle East are compared with typical "good" yields for open field crops in Table 28.

Table 28. Yields of vegetable crops grown hydroponically in desert greenhouses (CEA) and in open fields (OFA)

Crop	Hydroponic CEA[1] Yield/crop (MT/ha)	Hydroponic CEA[1] No. crops/year	Hydroponic CEA[1] Total Yield (MT/ha/year)	OFA[2] Total Yield (MT/ha/year)
Broccoli	32.5	3	97.5	10.5
Bush beans	11.5	4	46.0	6.0
Cabbage	57.5	3	172.5	30.0
Chinese Cabbage	50.0	4	200.0	—
Cucumber	250.0	3	750.0	30.0
Eggplant	28.0	2	56.0	20.0
Lettuce	31.3	10	313.0	52.0
Pepper	32.0	3	96.0	16.0
Tomato	187.5	2	375.0	100.0

[1] *Source: University of Arizona, unpublished results (1983).*
[2] *Source: Knott (1966)*

As the data indicate, the yield per crop is usually higher in greenhouses than in OFA because of the optimal growing conditions, balanced plant nutrients, etc., provided in controlled environments. Furthermore, depending on the vegetable grown, from 2 to 10 crops annually are possible with CEA, whereas only one crop per year is generally possible with OFA.

The more precise the control over the environment the more one is assured maximum potential yield. Tomatoes, for example, have exact growing temperatures for maximum yield - 15°C night, 23-25°C on sunny days and 17-20°C on cloudy days. Any deviation from these temperatures, will diminish the potential yield. Often during summer, in low desert regions, it is difficult to maintain 15°C night temperature due to the warmer outdoor temperatures. Temperatures above 15°C causes excessive respiration rates within the plant, therefore, lowering the yield. With absolute maintenance of proper growing temperatures, current yields on tomato crops may reach close to 500 MT/ha./year.

Optimum temperature control is made easier by locating the greenhouse at a proper altitude. To assure proper night temperatures during the hot late spring and early fall nights of the desert, greenhouses should be placed in the high desert regions. It is important to select an altitude that permits the desired degree of cooling during most times of the growing period. If evaporative cooling is used, a region of low humidity should be chosen for maximum efficiency. Even though it is easier to heat a greenhouse than to cool one, a grower must be careful not to locate at too high an altitude to avoid excessive heating costs during winter.

In Arizona, tomatoes should not be grown during the summer. Tomato prices are low during summer, as a result of open field tomato production in most of the United States, and day and night summer temperatures are high. The summer is a good time to remove the old crop, clean the facility and ready the greenhouse for the upcoming fall, winter, and spring production cycles.

If the growing media is free of disease and pests and the proper soil moisture and plant nutrition maintained, there will be little difference in the yield performance from one production system to another. Studies in Nova Scotia, Canada showed insignificant yield differences in tomatoes and cucumbers when using rockwool, perlite, and the nutrient film technique (NFT). Table 30 lists the results of these trials conducted in the spring of 1991. The rockwool was of the granular type. The plants of each system were placed in troughs with recirculated nutrient solutions. In the 1990 cucumber trials, perlite also produced a slightly greater yield than those plants growing in the NFT sys-

Tomatoes grown in a sand culture system with drip irrigation have exceeded 350 MT/ha./yr.

tem. Rockwool was not included in the 1990 Nova Scotia trial (Toms and MacPhail 1991).

Sand may be used as a growing media in desert regions, where organic materials are not plentiful and most conventional artificial media, such as rockwool, are costly to transport because of their low bulk density. Early tests at the University of Arizona indicate that crops grown in pure sand do as well as those grown in artificial media of peat moss and vermiculite or in various mixtures of sand and vermiculite, rice hulls, redwood bark, pine bark, perlite, or peat moss (Jensen 1975). Unlike many other media which undergo physical breakdown during use, sand is a permanent medium; it does not require replacement every 1 or 2 years. In comparing rockwool with sand culture, it has been demonstrated that each system will produce crops of excellent yield (Table 29); the difference in selecting a system lies in the technological capability of the

Table 29. Yield comparison between growing systems, Truro, Nova Scotia, Canada (Toms and MacPhail 1991)

TOMATO

Growing System	Yield Kg. per plt	Kg. per M^2	% Cull	Size gms/fruit
N.F.T.	6.15	18.45	6.6	220
Perlite Bags	6.05	18.15	7.3	208
Rockwool Bags	6.19	18.57	6.9	213

Cultivars (Ave): Trend, Capello, Seeded: Jan 10, Transplanted: March 1, Spacing: 3 plt/m^2, First Harvest: May 15, Terminated: August 16.

CUCUMBER

Growing System	Total Frt/Plt	Frt/Plt	Marketable Frt/m^2	%sm.	%med.	Size Grades %lg.	%e/lg	%cull
N.F.T.	51.4	44.7	62.6	19.7	49.1	17.1	1.1	13.0
Perlite Bags	53.7	47.3	66.5	18.4	49.3	19.7	1.0	11.5
Rockwool Bags	61.2	53.4	74.8	18.3	48.8	18.5	1.5	12.9

Cultivar (Ave.): Corona, Mustang, Seeded: Feb. 16, Transplanted: Mar. 21, Spacing: 1.4 plt/m^2, First Harvest: Apr. 26, Terminated: Aug. 30

grower and the economics of the production and distribution systems.

Table 30. Winter yield comparisons on tomatoes growing in rockwool and sand, Tucson, Arizona, USA

Treatment	Marketable Yield/plant(kg)	Ave. wt./ fruit(g)	Culls/ plant (kg)
Sand	.2	168.0	0.33
Grodan Rockwool	5.1	172.5	0.32
Domestic Rockwool	4.9	68.0	0.30

Cultivar: Goldsmith 203, <u>Seeded</u>: Oct. 3, 1985, <u>Transplanted</u>: Oct. 29, 1985, <u>First Harvest</u>: Jan 7, 1986, <u>Terminated</u>: Apr. 8, 1986.

Rockwool is replaced after every two to four crops, although a newly-developed rockwool from France can be used to produce up to eight crops. Some growers steam sterilize the rockwool each year by covering the growing beds with tarps and sterilizing the rockwool in place. Therefore, the number of crops grown in the rockwool is strictly up to the grower. In Europe, the rockwool is often returned to the manufacturer, reconstituted and reused for production.

Sand does not need to be replaced but must be fumigated once each year. If sand is used continuously for the production of one kind of vegetable, yields will gradually decrease over a period of years. This has been observed in tomato production and may be due to the gradual accumulation of root exudates. If sand is used, a crop rotation program is recommended. If this is not feasible, then a system in which the medium is periodically changed should be used, as is the case with rockwool.

Today, tomato yields from rockwool will approach 500 tonnes per hectare per year, or approximately 18 kg. per plant. This yield is from a single tomato crop with a harvest period of 7-8 months. Cucumber yields may exceed 700 tonnes/ha/yr; this yield is an accumulated yield from 2-3 crops over a period of one year.

In addition to the correct growing system, absolute control over the greenhouse environment and highly skilled management are necessary to produce high yields.

Any method of protected agriculture should first be put through a trial period, both horticulturally and economically, before it is recommended or adopted for use.

LABOR REQUIREMENTS AND SKILLS

Labor Intensity. The intensity of labor differs according to the system of protected agriculture used. Greenhouse vegetable production is much more labor intensive per unit of land than open field agriculture. The use of plastic mulch does not substantially increase labor requirements since, for the most part, mulch laying is mechanized. Removal of the mulch is generally done by hand, especially in smaller operations. This job can be laborious: vines covering the plastic can make it quite difficult to remove, especially when the edges are buried in the soil. On larger operations, non-degradable mulches can be removed with commercially designed equipment. If it is not, the plastic can be removed as in smaller operations by running a coulter

down the center of each row and picking it up from each side.

Row covers are usually installed over plastic mulch using a combination of mechanical and hand labor application. Equipment that will cover the rows in one operation is currently being developed. At the present time, however, farmer-designed equipment, used in conjunction with hand labor, is the most prevalent system. On small farms, the plastic covers are generally applied and removed by hand.

In greenhouse production, labor requirements for floricultural crops are quite similar to those for vegetables, except in bedding plant production. Bedding plants require about four persons/ha.; chrysanthemums require 11-12 persons. Poinsettia and Easter lily production are less labor-intensive than potted chrysanthemums, requiring only 6-7 persons/ha.

Greenhouse vegetable operations will normally employ 7-12 persons per hectare, depending on the management of the work program. To reduce on-site labor requirements, growers may contract with outside companies to perform a variety of cultural operations. Growers may belong to a local fruit packing cooperative. An individual grower who does not pack his fruit at the site of production requires less labor on his personal payroll. Often growers purchase their transplants and contract with pesticide applicators, often termed "custom applicators", for much of the pest control.

Normally, more labor is required in the production of tomatoes, lettuce, and cucumbers than in harvest. Production will require 57-65 percent of the labor needs while the remaining needs are attributed to harvesting and packing.

Cucumber production may require more labor than tomato production. Naturally, the higher the yields, the more labor needed to harvest and pack the fruit. Tomato size also affects labor needs: small cherry tomatoes require 12-15 times more picking labor than the large-fruited cultivars.

Increasingly, the packing and/or wrapping of certain vegetables is being automated. Even if not automated, mechanical aids are being incorporated into the packing process to increase labor efficiency and to expand the output of the facility.

Labor Skills. With limited training, a person can gain the necessary skills to work with plastic mulches and row covers. By comparison, the skill requirements for greenhouse systems of protected agriculture can be quite high, especially when fertilizer injector pumps, computers, etc. are included. Workers require both engineering and horticultural skills for maximum monetary return from such systems of high capital investment.

Training and Educational Needs. Most lesser-developed countries do not have personnel trained in protected agriculture — researchers, extension personnel and farmers. For success, it is imperative that all persons associated with protected agriculture have the necessary skills.

<u>Researchers</u>. The success of a program in protected agriculture starts with a research program. Countries wishing to establish any program of protected agriculture must invest the funds necessary to train their research personnel and establish research and demonstration programs. Research personnel must have a "hands on" work experience with protected agricultural systems if they are to develop a successful research and

development program for their own country. They may attend conferences and tour established programs in regions with similar climates and markets. Outside experts should be employed to assist in the design and operation of research facilities until a program is well-established, and the findings applied to the farming community. The research staff must consist of horticulturists, pathologists, entomologists, and soil scientists. Economists are also important to the research team: a given system of protected agriculture may perform well horticulturally, but be economically unviable. Marketing experts are critical to the program.

Extension Staff. Extension workers ensure the correct application of new technologies to the farm. Therefore they must not only be knowledgeable about the system being recommended, they must also have the "hands on" skills to demonstrate and assist interested farmers. To do this, they must work with the research staff to understand the systems being studied, the economic factors involved, and the methodologies of implementing the recommended systems on the farm.

A demonstration program must be established in combination with a research program. Extension workers will demonstrate new technologies during visits by farmers to the research and demonstration facility, and assist farmers in establishing new programs on the farm site. Descriptive literature should be available which details the methodology of protected agriculture; this information should be illustrated, especially in communities where farmers are not literate.

A successful program of protected agriculture is most successful when it is established by the government as a national priority, either through its own initiative or in partnership with a bank and/or lending agency or, on rare occasions, an agricultural products company. Because the profit margin in selling agricultural plastics is usually very low, such companies do not often finance and train personnel in protected agriculture and to markets.

Farmers. Without continuing help from extension and research personnel, farmers, especially small farmers, are generally unwilling to try new systems of protected agriculture. This reluctance is understandable: for many farmers failure means no financial resources for family support. Farmers will be more ready to try new technologies if they first plant a small trial area before making a large investment. They must also have the reassurance of access to authorities on protected agriculture.

CAPITAL REQUIREMENTS AND ECONOMICS OF PRODUCTION

The capital requirements differ greatly among the various systems of protected agriculture. Mulching is least expensive while greenhouses require the most capital per unit of land.

The fixed capital costs include land, fixed and mobile equipment, and grading, packing and office structures. The fixed capital costs for greenhouses clearly exceed those of other systems of protected agriculture, but vary in expense according to type of structure, and the environmental control and growing systems.

The operating (or variable) costs and fixed costs are annual expenditures. These can be substantial. Fixed costs include taxes and maintenance; operating costs include labor, fuel, utilities, farm chemicals, and packaging materials. Annual costs may correlate to some extend with capital investment: a more intensive culture is possible in a more advanced or intensive system of protected agriculture. Since systems of protected agriculture differ greatly, each system, and its capital requirements, will be discussed separately.

Plastic Mulch and Row Covers. Plastic mulch normally costs $385 to $585 per hectare, depending on the row spacing, type of mulch, and the supplier. Mulch which transmits only the IR radiation and not the visible light costs approximately 25 percent more. In estimating the capital requirements, the farmer must include the cost of the entire system as well as the mulch.

The University of New Hampshire, a leading institution in the United States on the use of plastics in agriculture, has conducted extensive research on the economics of protected agriculture. Table 31 shows the budget for the production of muskmelons in 1992. While the fixed and variable costs will differ by location and country, the budget in Table 32 serves as a model, listing the different cost categories, or sections, which must be included in any cost analysis when considering protected agriculture.

Table 30 shows that the material cost of the black plastic mulch was $585/ha. The cost to remove the plastic from the field was $140. Row covers would add an expense of $875-1,000/hectare; this includes the cost for the wire supports as well as plastic covers. In the United States applying and removing the row covers from the field was approximately $1250/ha. If floating row covers are used, the cost of materials might reach $1750-2,000/ha. Labor costs for applying and removing the floating covers are less than for polyethylene. Material costs for floating row covers would be reduced if they were reused, which often happens. Polyethylene covers are not reused.

Table 31. Muskmelon - costs and returns of production (1992 New Hampshire Cooperative Ext. veg. crop budget)

Section	$/ha (U.S.)	Section	$/ha (U.S.)
I. FIXED COSTS		III. MARKETING COSTS	
Machine Insurance,		Retail	$9,375.00
Housing, Taxes	$ 50.00	Wholesale	——
Machine Capital		**Total**	**$9,375.00**
Recovery &			
Interest	567.50		
Management	1,875.00	IV. TOTAL COSTS	
Land Charge & Taxes	375.00	Without labor	
Total	**$2,867.50**	and management	$14,140.50
		With labor and	
II. VARIABLE COSTS		management	18,325.50
Seed	$ 225.00		
Rye Seed	37.50	V. GROSS RETURNS	
Fertilizer	236.25	Retail	
Lime, custom applied	50.00	(34.0 tonnes/ha.)	37,000.00
Black Plastic	585.00	Wholesale	——
Herbicide	157.00	**Total**	**$37,000.00**
Insecticide	33.50		
Harvest Supplies	1,125.00	VI. RETURNS BY CATEGORY	
Misc. (remove		Gross Returns	$37,000.00
plastic, 20 hr.)	140.00	Returns Minus	
Fuel and Oil	80.00	Market Costs	28,125.00
Machine Repairs	145.00	Returns Minus	
Hard Labor - plant,		Market and	
weeding (80 hr.)	560.00	Variable Costs	22,042.00
Hand Labor - harvest		Returns Minus All	
(250 hr.)	1,750.00	Costs (Profit)	19,124.50
Machine Labor			
(50 hr.)	467.50	Returns to Labor	
Operating Interest	491.25	and Management	23,359.50
Total	**$6,083.00**		

Source: Sciabarrasi and Wells (1992).

Using row covers often results in an early harvest, which normally bring a benefit in higher market prices. Without this benefit, it may not be economically advantageous to use row covers or plastic mulch.

In Connecticut, farmers use floating row covers to obtain early harvest in late July and August, when the retail price of red or yellow peppers remains about $5/kg. These prices drop to $1.25 to $2.50/kg. in September when local production reaches its peak. Thus, the economic benefit of floating row covers is correlated with production of ripe colored peppers in July and August. An estimate of increased gross income was obtained by multiplying the difference in retail price between August and September, $2.50/kg., by the difference in marketable yield as of mid-August, between peppers planted early and grown under row covers and those grown conventionally (Gent 1989). The benefit of early harvest is possible, however, only through an additional investment in sprinklers — a cost which must be added to the calculations. Floating row covers do not provide good frost protection which may be needed for seedlings planted in April, before the frost-free period. However, sprinkling throughout the night or covering overnight with a second impermeable cover will provide frost protection. Using sprinkler irrigation for frost protection will increase the cost of production by an estimated $1,250/ha. With these precautions, Connecticut farmers can produce a crop of peppers starting as early as late July, continuing until a killing frost in the fall.

In California, where over 3200 ha. of vegetables are devoted to plastic-covered trenches, the 1989 costs for 25 micron polyethylene film was about $260/ha. Labor costs to lay the plastic trench were $75/ha and to remove the plastic the costs were about $45/ha, plus a fee for disposal (Mayberry 1988). In addition to providing earlier melon harvest, the plastic barrier also offers early season insect exclusion, delaying the onset of virus transmitted by aphids.

Greenhouses. While greenhouse production systems may be far more expensive than open field systems of equal land area, open field systems of protected agriculture are normally more extensive in field area than in greenhouse production. Therefore, capital and operational costs for each growing operation may be quite similar.

Open Field versus Greenhouse Culture. A typical greenhouse operation is one hectare; a farming operation, using polyethylene mulch and row covers, typically total at least 40 hectares. The operating cost for a one hectare greenhouse growing cucumbers may reach more than $400,000 per year; the operational cost for the 40 ha open field operation, growing a similar crop or a variety of vegetable crops, may also total $400,000. The open field costs are for one season only; the greenhouse costs are for a year.

These farming methods are equally expensive to operate and may even be similar in capital investment. Depending on the location and cost of land, a one hectare greenhouse may equal the cost of 40 hectares for open field agriculture plus the support facilities and equipment. If the capital cost is included as depreciation in the overall operational cost, the $400,000 to operate a one hectare greenhouse or a 40 ha. open field operation, may reach nearly $500,000/ha., when taxes and debt service are added in.

The net return on the investment may also be similar, but the risks may be different. Cost per unit of output is much higher from a greenhouse than from open field agriculture. Crop failure in a greenhouse can cause greater financial suffering than a similar loss in the field.

There are many factors involved in selecting a system. Sometimes land costs are prohibitive and the only option, economically, is a greenhouse facility. This certainly is the case in The Netherlands. Greenhouse costs also vary according to type. Single unit, high tunnel polyethylene structures used seasonally are far less expensive than the large gutter-connected glass structures with sophisticated environmental control systems used year-around.

High Tunnel Polyethylene Structures. High, narrow polyethylene greenhouses, measuring approximately 4.3m x 29.3m are becoming increasingly popular, especially in Europe, the Middle East, and Asia. They are often termed hoop houses, walk-in tunnels, or unheated greenhouses (Wells 1991). While not precisely greenhouses, they operate on the same principle.

The quonset-shaped structures are made of metal bows connected to pipes that are driven into the ground and covered with clear plastic (Wells 1991). The ends are enclosed; there is no supplemental heat or mechanical ventilation. Water is provided by trickle irrigation. Ventilation is accomplished by rolling up or lifting the polyethlene on the sides of the greenhouse. Since the structure is used primarily in the early spring to provide an early harvest, the film can be rolled up to the ridge of the greenhouse roof in summer, covered with black plastic to prevent degradation of the plastic by the sun, and stored there until the following season.

Trials conducted at the University of New Hampshire (Wells 1991) show the cost of construction to be approximately $1700 for a 126m^2 structure ($13.48/m^2) including the plastic cover, black plastic mulch on the floor

and trickle irrigation. Table 32 lists the economic analysis, compiled by Wells and Sciabarrasi (1991), of the high tunnel tomato production system.

Wells (1991) was able to advance the planting dates by three to four weeks. If the planting dates are earlier, two layers of plastic should be placed over the structure to contain in more heat during cold periods. A portable heater could be also installed, but at this point a grower would have a greenhouse rather than an inexpensive high tunnel (Wells 1991). Table 31 shows a good return per 126m^2. If the tunnels totaled one hectare, the net return would be nearly $113,000. The selling price of the tomato fruit, $3.52/kg, in Table 31 is excellent; but this may not be the case in other regions of the United States at this time of year. A grower must investigate the market before committing any large sums of money to any method of protected agriculture.

Controlled Environment Greenhouses. Greenhouses are expensive, especially if the environment is controlled through the use of heaters, fan and pad cooling systems and computer controls.

The fixed capital, production costs and returns are outlined in Table 32. The costs are for greenhouse cucumber production and tomato production are quite similar with regional variations. In the last 20 years, greenhouse construction and operation costs have risen sharply. In 1972, the cost to build a one ha. greenhouse for tomato production in California was approximately $275,000. The total production and marketing cost of such a unit was about $42,000 (Dalrymple, 1973). Today, the construction cost is three-five times greater, with some glasshouse ranges approaching $140/m^2.

Since 1972, fixed capital costs, labor and energy have all risen substantially, causing a dramatic increase in the operational costs of a new greenhouse. When labor reaches $10 U.S. or more per hour, growers the world over turn to mechanization.

During the past twenty years also, new vegetable cultivars have been improved. These cultivars, used with new growing

In Arizona, cucumbers in sand culture are grown in greenhouses where the environment is controlled through the use of natural gas heaters and fan and pad evaporative cooling.

Table 32. Economic feasibility of high tunnel tomato production, 1992[1]

I. Structure Costs (amortized over 10 years)			
A. Frame (posts and bows)	$ 800.00		
B. Side Boards	130.56		
C. Pipe & T Handles (roll-up)	137.50		
D. End Walls	22.56		
E. Construction Labor			
32 hrs. @ $8/hr.	256.00	Per tunnel	Per m^2
		$1,346.62	$10.70
II. Plastic and Trickle Irrigation (amortized over 5 years)			
A. Cover, I layer, 150 micron			
7.3 x 30.5m	203.79		
B. Black poly sheeting,			
150 micron 5.0 x 30.5m	63.10		
C. Trickle Irrigation	85.00	Per tunnel	Per m^2
		$351.89	$2.79
III. "Annual" Expenses and Returns		Per tunnel	Per kg.
Receipts 910 kg @ $3.52/kg.		$3,200.00	$3.52
Marketing Costs (25% of receipts)		800.00	0.88
Production Expenses			
A. Plants, 0.65m^2/plt,			
192 plts/ @ $0.15	$ 28.80		
B. Stakes, 192 @ $0.25	48.00		
C. String, 762 m	15.00		
D. Fertilizer - starter,10-20-20	5.00		
E. Fertilizer through			
trickle irrigation	20.00		
F. Containers, 100 9 kg. boxes			
@ $1.10 each	110.00		
G. Labor @ $8.00/hr.			
Till, spread fertilizer,			
and plant, 3 hrs.	24.00		
String plants, 3 hrs.	24.00		
Prune, 2.5 hrs.	20.00		
Harvest, 8 har x 4 hrs.	256.00		
Annual maintenance, 10 hrs.	80.00		
H. Misc. (small tools, repairs,			
rototill)	25.00		
I. Operating Interest @ 5%			
$656 (A thru H)	32.70		
Subtotal Variable			
Expense (A thru I)	$ 688.59		
J. Annual Capital Recovery			
and Interest			
Structure ($1,346.52,			
10 years, 11%)	228.66		
Plastic & Irrig. ($351.89,			
5 yrs,11%)	59.75		
Subtotal Fixed Costs	**$ 288.41**		
Variable and Fixed Production Costs		$977.00	$1.08
Net Returns (Receipts-Mkt.Costs-Prod.Costs)		**$1,423.00**	**$1.56**

[1]One tunnel @ 4.3m x 29.3m = 126m^2

Source: Otho Wells and Mike Sciabarrasi, University of New Hampshire, Durham, N.H., USA, Personal communication

systems, better environment control, and computer management, have helped to nearly double both tomato and cucumber yields.

High tunnels are used seasonally; controlled environment greenhouses are normally in use year-around. Tomato yield from high tunnels, as illustrated in Table 31, are approximately 72 tonnes/ha, an excellent yield considering that there is little control over the environment and so little investment. Controlling the environment during the entire year is costly but results in high yields, especially in the high light regions, where tomato yields may sometimes approach 500 tonnes/ha.

Floricultural crops are generally grown in the United States in the more expensive environment-controlled greenhouses because of the fluctuations in climate. In Colombia, flowers are grown in greenhouses that are basically only rain shelters. If a cooler temperature is desired, the growing units are simply built at a higher altitude. This is possible because the day to day temperatures in Colombia are relatively constant.

While the fixed capital costs for greenhouse floricultural production are similar to those of vegetable crops, the fixed and variable operational costs are very different.

Because production costs, returns and recommendations for each floricultural crop vary so greatly, persons interested in floricultural crops are advised to contact their agricultural university or a regional company representative knowledgeable in the specific flower crop. Recommendations will depend on climatic region, among other considerations. An excellent series of publications on floricultural crops is available from the Ohio Cooperative Extension Service, The Ohio State University, Columbus, Ohio. These publications cover nearly all aspects of potted Easter lilies, poinsettias, bedding plants, chrysanthemums, and the growing of perennial and biennial floricultural crops.

While the recommendations for a given crop grown in Ohio may be different than for another region, the Ohio publications will provide a valuable, basic knowledge of this intensive agricultural industry.

Costs, Prices, and Profits. It is clear that greenhouses may require a lot of capital. Dalrymple (1973) quotes Liberty Hyde Bailey that as early as 1897 the cost structure of greenhouses was recognized as being high:

"The person who desires to grow vegetables under glass for market must, first of all, count up the costs and the risks.

- Glass houses are expensive and they demand constant attention to repairs.

- The heating is the largest single item of outlay in maintaining the establishment. Moreover, it is an item upon which it is impossible to economize by means of reducing temperature, for a reduction of temperature means delayed maturity of the crop.

- Labor is the second great item of expense.....This, however, may be economized if the proprietor is willing to lengthen his own hours, but economy which proceeds so far that each one of the plants does not receive the very best of care is ruinous in the end."

Since Bailey's time, wages have risen more than heating costs. In Table 32, under variable costs, wages account for

$175,000 of the $319,000. While these figures represent a production system in the Arizona desert, wages would still far exceed heating cost, even in the colder northern regions of the United States. Wages are certainly the greatest single expense in greenhouse production, followed by amortization costs and then energy costs, especially where heating is necessary. In Table 32, about two-fifths of the expenses are fixed costs and about three-fifths are variable (operating) costs. Depreciation and interest on investment accounts for most of the fixed costs.

Gross returns from greenhouse vegetables must be high. This is accomplished by high prices for the product and/or high yields. There is little room for error; therefore, it is imperative that there are no shortcuts in environmental control, competent management, or any other factor of production. Today, in the United States, retailers commonly double their sale prices over the wholesale price to the grower. Such a high mark-up can cause great consumer resistance. In Europe the typical mark-up is a more modest 30-40 percent. In Ohio in 1961 (Dalrymple 1973), it was common to have a retail margin of only 30-32 percent, and greenhouse tomatoes in the United States were usually of much greater value than open field products. Today this is not necessarily true. Quality control, packaging and transportation of open field tomatoes has improved, and, new varieties are better able to withstand shipment. At the same time, greenhouse tomatoes and peppers are being imported into the United States from Holland; Dutch growers package tomatoes of absolutely uniform size and color in packaging cups thus safeguarding each tomato against bruising during shipment. This new marketing strategy will be further discussed in Chapter XII. Such new strategies better the position of the greenhouse vegetable industry to compete in the marketplace.

The seasonal variation in prices presents a tremendous economic challenge in the production of greenhouse crops during a period of high demand. The greenhouse grower is very vulnerable to price changes: he does not have much short-term production flexibility when faced with low prices. High overhead costs mean he cannot easily let his house lie idle during periods of low demand. Growers are constantly working on ways to reduce operating costs in order to increase profits, such as reducing labor costs by using bees to pollinate tomatoes for fruit set rather than labor with hand pollinators, or by insulating a greenhouse against heat loss.

Conditions Influencing Returns. A number of variables which may not show up in the yearly financial balance sheet influence the returns to greenhouse operators: economies of scale, physical facilities, cropping patterns, and government incentives (Dalrymple 1973.)

Economies of Scale. The size of any system of protected agriculture will depend on the market objectives of the farmer. Most protected agricultural endeavors are family operated. Often the products are retailed directly to the consumer through a roadside market at the farmsite. Farmers commonly use several systems of protected agriculture, both greenhouse and open field operations where crops are mulched and covered with row covers. Farmers also diversify their crops in order to offer an array of products for sale at a farmsite market.

In the developed world, greenhouse operations tend to be of

Table 33. Greenhouse cucumbers (1 ha.) - estimated capital, operating costs and returns (Tucson, Arizona 1991 - 3 crops)

I. Fixed Capital Costs	Cost/ha	Annual Production Cost & Returns	
Greenhouse			
Structure (polyeth. roof, polycarbonate sides and ends)	$193,578	II. Gross Sales	
		Boxes of Cucumbers	
		(16 cucs/box)	87,500
Heating and Ventilation	176,778	Gross Revenue	
Elect. and Plumbing	62,000	$9.50/box	$831,250
Sand Culture	57,500		
Irrig. and Fert. Equip.	24,000	III. Variable Costs	
Construction Labor	75,000	Nursery[1]	28,818
Sub-Total	**$588,856**	Greenhouse[1]	122,817
		Packing[2]	117,792
Service Building		Energy (Heat/Cool)	49,363
Structure	63,960	**Sub-Total**	**$318,790**
Cold Storage	40,000		
Packing Equipment	15,500	IV. Market Costs	
Standby Generator	30,000	20% of receipts	$166,250
Office & Restrooms	5,000		
Shop Equipment	5,000	V. Fixed Costs	
Pesticide Equipment	5,000	Roof Cover Replac.	
Sub-Total	**164,460**	(every 2 years)	8,000
		Repairs & Maintenance 15,000	
Vehicles		Office	13,600
Trucks, etc.	46,500	**Sub-Total**	**$ 36,600**
Ancillary Systems		Total Costs	
Well, fuel storage, roads, pipe and wiring, runs, sewers, fencing, etc.	90,000	(III+IV+V)	$521,640
		Net Operation	
		Profit (NOP)	309,610
		Less interest	-31,000
		Less depreciation	
Total	**$889,816**	(10 years)	-90,000
	($88.69/m^2)	Taxable Income	
		(NOP-Int.-Depr.)	$188,610
		Income Tax	56,583
		Debt Service	
		(10 years)	90,000
		Post-Tax	
		Net Return	$163,027

[1] Includes labor $7/hr, plant propagation, pest control, fertilizers, etc.
[2] Includes labor $7/hr, boxes and shrink film.

a size that can be operated by one family — from 4,000 to 8,000m^2. In Canada, for instance, greenhouses rarely exceed 4,000m^2 (Dalrymple 1973). In Holland, where particular emphasis is placed on labor efficiency, the average heated greenhouse holding is about 8,000m^2. A unit of 4,000m^2 can be operated by two to three laborers, with additional help at periods of peak activity. This amount of labor can usually be provided by the owner and his family. Moreover, the owner will pay close attention to management - a most important factor.

Diseconomies of scale in cooperative greenhouses mainly result from inefficient decision-making, (Dalrymple 1973). Since labor and energy account for a large proportion of annual expenses, these costs remain generally proportionate to the increased size of the operation. Indeed, labor costs may rise significantly if it is necessary to recruit labor outside the family. Greenhouse owners who hire a highly qualified manager may have to operate a larger than family-size greenhouse in order to offset the expert manager's salary.

Economies might accrue with increased size when (1) there is a unique opportunity to mechanize certain operations, (2) labor can be more efficiently utilized, (3) low cost capital is available, (4) there are economies in the purchase of packaging materials and in marketing, or (5) some special management skills are available (Dalrymple 1973). Another consideration is the size of the greenhouse in relation to fixed capital cost. Economies of scale may occur where larger greenhouse units cost less per m^2 than smaller units. A single four hectare greenhouse under one roof may cost less per m^2 than a four hectare greenhouse complex, composed of four-one hectare units, because of fewer sidewalls in the single unit.

Market demand may require the greenhouse to be larger than a family unit. Buyers of horticultural produce often deal with packing houses because they want greater quantities of a product than is available from a single family unit. However, packing houses are cooperatives made up of many growers, each with a different cultural program and possibly differences in fruit or flower quality. It is therefore an advantage to buy from one large producer promising a reliable product of consistent quality. Such large greenhouses are being constructed in the Southwestern part of the United States and in Northern Mexico; there growers are building extensive greenhouses in high light regions to produce high quality, large quantity of crops throughout the year.

In Eastern Europe, large units were built more for ideological commitment to large scale agriculture than for economic reasons.

As technological advances in greenhouse agriculture continue, family-sized greenhouse units are getting larger. The average size of Dutch vegetable houses has increased substantially; the optimum-sized, three person unit, is now one hectare. Similar modest increases in size will continue in Holland and in other countries (Dalrymple 1973).

Physical Facilities and Location. Another variable that influences the return to greenhouse operators is intensity of production.

Intensity can be determined by the type of structure and the degree of environmental control equipment. When a grower constructs a new house, should it be complex or made as simple as possible? Should it be a high tunnel greenhouse ($13.50/m^2) or a controlled environment greenhouse ($88.70/m^2)? The major factors for consideration include the availability and cost of capital, land, construction and operating costs, climatic factors during the anticipated production season and the existing price structure.

The high tunnel growing unit, a simple structure with little environmental control equipment, is less expensive to build. Selecting such a unit reduces the amount of capital needed and the amortization costs.

The more complex structure, with a more complete environmental control system makes year-around production and early harvests possible, thus enabling the grower to realize the higher prices of earlier or late markets, and/or to increase the frequency of cropping. Year round production offers year round employment of laborers who, over time, will become highly skilled and of great value to the success and continuity of the business.

As Tables 32 and 33 demonstrate, the environmentally controlled greenhouse produced only one-third more revenue than the high-tunnel structure. This differential is slight, especially when comparing the construction costs on a m^2 basis. Other common considerations in selecting a greenhouse are site and market specific, as well as the personal preference and goals of the farmer.

The various factors influencing selection of greenhouse type

also influence the regional location of greenhouse production. In the past, greenhouses were built near urban areas, in both Europe and the north-central and northeastern United States. In China, large greenhouse operations are located within the city boundaries.

With improved roads and transportation facilities it is less important to locate greenhouses close to large population centers. In recent years, new greenhouse construction has been more widely diffused throughout the United States. The newer areas of production combine at least several of the following factors, which contribute to lower costs (Dalrymple 1973):

- high sunlight intensity undiminished by air pollution,
- mild winter temperatures,
- infrequent violent weather (tornadoes, high winds, hail, excessive snow),
- low humidity during the summer for air cooling,
- a good water supply, low in salt.

Other desirable factors are cheap fuel, such as natural gas (not bottled propane), cheap electricity, availability of labor with agricultural experience, and low taxes. In the United States, many of these above factors are found in New Mexico, Arizona, Nevada and California.

The situation is similar in Europe; much of the new construction is in southern Europe, where the weather is milder. Most of the new structures are of plastic as they are in the United States. The greenhouses built in southern Europe are very inexpensive, with little or no environmental control. Most are in Italy — especially Sicily, and southern Spain. A large part of the production is shipped to northern Europe. In England there is no one low-cost area of production. In the 1960's and 70's, there was a pronounced expansion of greenhouses in eastern Europe.

Cropping Patterns. The type of greenhouse structure will have bearing on the cropping pattern. In the United States, a high tunnel structure or any structure not fitted with environmental control equipment for heating and cooling will be used only on a seasonal basis. This is true for greenhouses in most of the world, with the exception of those in the more tropical regions such as in Colombia, Singapore, etc. The more elaborate facilities, which provide some control over the environment, make multiple cropping possible. The increased cropping alternatives in turn make possible more choice in crop selection.

The question of which are the most economical cropping patterns for any given region of the world has not received a great deal of attention. In the United States there have been few alternatives in cropping patterns, especially in greenhouse vegetable production. In the U.S., for reasons of economics, it is common for greenhouse vegetable operations to change to flower production, especially in structures with more elaborate environmental control systems.

Growers throughout the world are currently experimenting with alternative crops, such as herbs, in the United States, and Norway, and bell peppers in a variety of colors, in Holland.

Each cropping pattern requires different management skills and resources for each farm. As eating habits change — especially in the United States where consumers are increasingly conscious of diet and the nutritional value of fruits and vegetables — growers must continually look for alternatives to existing cropping patterns.

d. Government Incentives. Financial returns may be influenced by government action. Governments in England, Ireland, and France have stimulated the greenhouse industry by providing grants or low interest loans toward construction costs (Dalrymple 1973). In Holland, the price of fuel was subsidized by the government.

In China, government assistance has led to a rapid increase in protected agriculture and, in turn, food stability.

The government assists growers by providing a subsidy on the cost of plastic for mulches, row covers, and greenhouses. In countries where quality plastic is not manufactured domestically, growers face high import tariffs on plastic mulches. Without subsidies, the cost of such mulches is prohibitive, and agricultural plastic has not been used to increase food production.

The attitudes of farmers toward investments in protected agriculture, with or without government incentive programs is not fully understood. Exactly why governments provide incentives is not always clear. It may be just a way to stimulate agriculture or the government may be concerned with reducing imports or exports, or other factors, such as political influence. What is understood is the result of governments placing high tariffs on the importation of agricultural products, such as plastics, on the growth of protected agriculture. The growth results are nil.

12

MARKETING AND DISTRIBUTION*

The process of marketing and distribution is a key factor in determining the success of any system of protected agriculture. One reason for this is the special nature of protected agricultural crops.

Most crops are of high value and perishable. Often termed luxury crops, they are highly regarded for the variety they lend to meals in the form of vitamins and minerals, and the beauty provided by floricultural crops. Few crops produced by protected agriculture provide the basic food ingredients for human survival, as do cereals. An exception to this is the use in China of plastic mulch to produce not only horticultural crops but also agronomic crops such as cotton, peanuts, etc. How-ever, plastic mulch, row covers, and greenhouses usually are used to produce high value/perishable food and floricultural crops when local field production is not readily available.

The added ex-pense of using any of the protected agricultural systems results in higher retail costs. Any development which lowers the cost of products from protected agricultural systems or increases the consumer's level of income encourages purchases of these products (Dalrymple 1973).

Greenhouse tomatoes are packed into containers with individual pockets to prevent severe bruising of the fruit in shipping.

INTRA-NATIONAL COMPETITION

Protected agriculture faces a wider range of intra-national competition from domestic production than may be evident. Major competition comes from both fresh and processed products grown in open field production or by other protected agricultural systems.

Field Production. Open field production is the most serious competition, particularly in countries with a wide range of climates. This is especially the case with food crops. Because floricultural crops are most often grown in some type of structure, competition from open field production is not a significant market factor.

Protected agriculture, especially greenhouse food crop pro-

duction, rarely competes with local field production of the same crop during the same period of time unless 1) the crops from systems of protected agriculture have a market quality advantage or 2) field production is a marginal system of production due to climatic conditions.

Producing food crops of better quality with systems of protected agriculture provides few benefits over field production, except for floricultural crops. Although some U.S. hydroponic vegetable growers have claimed quality advantages of tomatoes and lettuce, this superiority has not been demonstrated. Through improved cultivars and packaging techniques, open field producers can market an excellent quality product.

The situation is different for greenhouse cucumbers. As discussed earlier, producers in North America today are growing the long, seedless cultivars from Europe, which are distinctly different from the short, open field seeded types. The European cucumbers have not been grown successfully outdoors since the cucumber abrasive leaves damage the fruit during periods of wind and pollination by bees is not wanted. Many consumers strongly prefer greenhouse cucumbers because of quality differences.

In the most northern countries of Europe, there is little or no field production of tomatoes and cucumbers because of cool weather conditions. In these regions, greenhouse production does not face competition from local field production, except for competition from products imported from southern Europe. In France and Great Britain, row covers are commonly used for many vegetable crops and compete with similar greenhouse products.

Systems of protected agriculture may compete with each other. For example, early market tomatoes produced in the field through use of plastic mulch and row covers may compete directly with greenhouse tomatoes.

Since protected agricultural systems are used, for the most part, to produce crops out of the local season, the major form

*For more up-to-date information see atttached Annex on "Marketing and Distribution of Protected Crops" by Alan J. Malter.

Tomatoes of uniform size and color are sent to markets with literature identifying the fruit as being greenhouse grown.

transportation system, and the rapid escalation of fuel costs in 1972, due to the Arab oil embargo, which increased the cost of greenhouse heating, with natural gas, 5-6 times.

For these reasons, the southern regions of the United States became a major source of fresh produce to the north. While the cultural energy inputs are similar for both open field and greenhouse agriculture, the energy input for heating and cooling greenhouses is far different (Table 34).

It takes nearly 35,000 more Kcal/kg. of tomatoes to grow in a conventional Ohio greenhouse over the open field. New energy saving technology might greatly reduce the heating and cooling energy inputs for greenhouses, but until now this has not occurred to the degree desired.

of competition usually comes from imported produce. In the northern hemisphere, the imported produce has come from the more southerly regions in the same country or continent. The reverse would, of course, be true in the southern hemisphere.

When the first greenhouses were built in the northern United States, good interstate transportation did not exist. Since the construction of excellent interstate freeways and the development of large semi-trailer trucks capable of hauling heavy loads over long distances, most northern greenhouse producers have been unable to compete with produce shipped in from the southern hemisphere. In 1965 Ohio had more than 240 ha. of glasshouses used for vegetable production, which accounted for 75 percent of the total U.S. production. In 1982, only 64 ha. remained, accounting for only 25 percent. The decline came for two reasons: The much improved interstate

Table 34. Total energy inputs - MKcal/tonne[1]

Energy Input	Open Field (Ohio)		Greenhouse (Ohio)	
	Processed	Fresh	Conventional	Possible
Cultural Energy	0.28	0.29	0.20	0.15
Processing and Containers	2.27	0.60	0.50	0.50
Storage	0.13	—	—	—
Transportation	1.26	2.77	—	—
Heating and Cooling	—	—	35.28	4.41
Total MKcal/tonne	3.94	3.66	35.98	5.06
Kcal/kg. of tomatoes	4,334	4,026	39,578	5,566

[1]*Taken in part from paper entitled, "Energy and the Food Chain" by Warren Roller, Ohio Agricultural Research and Development Center.*

Possibly the most interesting comparison is to look at the energy requirements to ship horticultural commodities produced in the southern latitudes to the north. In Table 35, the Kcal required to ship one kilogram of tomatoes 2,000 km. is given. Shipping one kg. 5,000 km north by semi-truck, expends approximately 1865 Kcal of energy. Growing one kg. of tomatoes in a northern region takes nearly 40,000 Kcal of heat energy. In other words, during the winter it is much cheaper to grow tomatoes in a southern latitude, in an open field, and distribute the produce in the north.

In the United States, it is common for horticultural products to be shipped across the continent.

Table 35. Energy use by mode of transportation[1]

Mode	Kcal/Kg.tomatoes/2000 km
Rail	236
Semi Truck	746
Air	4542

[1] Source: U.S. Dept. of Transportation, "A Prospectus for Change in the Freight Railroad Industry", p. 106, 1978.

A variety of greenhouse vegetables, packaged to extend the shelf life, on display in a food market.

Vegetable cultivars developed for shipping and new transportation techniques will continue to give consumers in the northern hemisphere horticultural commodities as good or better than those grown locally in a greenhouse.

The differences between open field and greenhouse crops, with the exception of cucumbers and floricultural crops, may add up to only subtle visual differences at the retail store. Greenhouse products are seldom labeled, nor can they be distinguished by packaging .

In lesser developed countries, internal transportation systems are often poor, thus making long distance shipping of quality perishable products difficult or impossible. In countries having a wide range of climates, like China, protected agriculture in the northern regions is making a major impact on the production of out-of-season crops, as was the situation years ago in the United States.

Competition from processed products grown by means of protected agriculture is virtually non-existent. Nearly all food crops produced by means of protected agriculture are consumed in fresh form, except gherkins, or pickling cucumbers, in Europe. As production expands into other crops which have processed counterparts, such as frozen strawberries, this issue of competition could become of greater importance.

Table 36. Comparative greenhouse yields and energy usage

State	Vegetable	Tonnes/ha/yr.	Energy Input Kcal/kg.
Arizona	Cucumber	525	6,855
	Tomato	300	11,997
Ohio	Tomato	300	39,578

Protected Agriculture Production. As mentioned earlier in this chapter, different systems of protected agriculture may compete, such as early spring crops produced under row covers competing with similar crops grown in greenhouses. Competition may also come from greenhouse crops grown at different locations within a country.

As previously discussed, in the winter, the energy requirements to heat a greenhouse in the northern region are far greater than the energy used for greenhouse production in warmer southern regions. In the northern region, most of the energy is used for heat; in southern latitudes, the energy is used for fan and pad evaporative cooling, as well as heating.

Table 36 compares energy usage between Arizona, in the arid south, and Ohio, in the north. The energy to grow a kg. of tomatoes in Ohio is more than three times greater than that energy needed in Arizona.

INTERNATIONAL COMPETITION

Undoubtedly, the greatest competition for protected agriculture in many countries

Along with extending the shelf life of the lettuce, the bag identifies the grower and method of production.

comes from imported produce, both greenhouse and field grown.

The percentage of exported fruits and vegetables that are products of protected agriculture is unknown for today's market. In 1983-85, horticultural products constituted about 12 percent of world agricultural trade (Islam 1990). The developing countries accounted for 37 percent of these exports, with an absolute value of US $9 billion. Most produce was imported by the developed countries, which consumed 83 percent of world imports of horticultural products during 1983-85. The United States and Japan were the fastest growing import markets during 1975-85. The developing countries share in this import market, increased from 14 percent in 1975-77 to 17 percent by 1985. Western Europe remains the largest import market, about 80% larger than the United States market.

Most world trade in horticultural products was among the developed countries. During the early 1980's, 72 percent of their horticultural imports came from other developed countries; 80 percent of their produce was exported to these countries. The developing countries send about 20 percent of their exports to other developing countries. Since 1975, fruits have constituted the major portion — some 70 percent — total horticultural exports by developing countries. The growth of import demand, especially in the developed countries, will determine the future of horticultural exports (fruits and vegetables) of developing countries (Islam 1990).

World trade in floricultural crops, such as cut flowers, reached U.S. $1.9 billion in 1987, a 17 percent increase over the previous year (Morrow 1989). The Netherlands, the world's largest flower supplier, produced more than 71 percent of all exports. Colombia, the second largest exporter, lagged behind with 9 percent. The two major markets for cut flowers are Europe and the United States; Europe imported 80 percent of the total import value in 1987 and the U.S. 15 percent. Japan is the dominant importer in Asia, a growing third market, with 3 percent of cut flower imports in 1987.

Each of the three major markets is dominated by a different mix of suppliers. The major suppliers of the European market are the Netherlands, Israel, Italy, and Spain, which account for 88 percent of the imports. Colombia and the Netherlands dominate the U.S. market with 82 percent. The Netherlands, Thailand, Taiwan, Malaysia, and New Zealand supply a combined total of 80 percent of flowers imported by Asia.

Colombia, Thailand, and Kenya were, in 1987, the only developing countries to have market shares of 1 percent or more of global cut flower exports. Mexico and Turkey may become more important exporters of cut flowers in the near future.

Future demand for cut flowers is uncertain. Industry experts predict strong growth in the U.S. market, which still demonstrates significantly lower per capita consumption than many European countries.

Statistics for cut decorate greens, live plants and bulbs are similar to those for cut flowers. The Netherlands is also by far the greatest producer of this floricultural crop (Johnson 1990).

Issues of trade and competition are different in North America than in Europe; therefore the two regions will be discussed separately. More comprehensive and updated information is provided by Alan J. Malter, in the Annex to this publication.

North America. Mexico provides major competition in food crops in the United States and Canada. In the United States, more than over 50 percent of the vegetables consumed during winter come from Mexico. In 1987, 98 percent of all the fresh market tomatoes imported into the United States and 96 percent of the cucumbers came from Mexico.

The percentages of vegetables on the American market which were grown in Mexico are:

Cucumbers	43%	Cantaloupe	15%
Eggplant	33%	Watermelon	4%
Peppers	28%	Honeydew	13%
Tomatoes	22%	Asparagus	9%

The American imports of Mexican fresh market tomatoes increased from 165,000 tonnes in 1963-64 to 291,000 tonnes in 1969-70 (Dalrymple 1973). In 1987, the tonnage was approximately 400,000 tonnes.

Some American consumers think that domestic greenhouse tomatoes are of finer quality than Mexican imports and are willing to pay higher prices for them. Because of lower production costs in Mexico — labor, fuel, and water — U.S. greenhouse growers have had difficulty in competing with the Mexican imports. In his 1973 book on greenhouse vegetables, Dalrymple noted the rapid increase in acreage of greenhouses in the American Southwest, a region, he felt, that might suffer the most severe competition from Mexico. His warning was correct: all the greenhouses built in that region in the 1970's are no longer in operation.

Today, Mexican growers are rapidly employing the technologies of protected agriculture, mainly plastic covered trenches, for early crop production, as well as greenhouses for both cucumber and cut flowers — predominantly roses — production. In time, Mexico will also export potted floricultural crops, once quarantines to prevent soils from entering the U.S. from Mexico are lifted. One day, potted plants with artificial soils will be permitted into the United States as they are presently from Canada. The present quarantine will certainly be abolished when the upcoming free trade agreement between Mexico, Canada, and the United States is concluded.

Paralleling the new interest in Mexico in growing greenhouse vegetables is a renewed interest in growing greenhouse vegetables in the American Southwest. Owners of greenhouse vegetable operations in the north are building operations in the Southwest in order to supply preferred buyers on a continuous basis.

With use by Mexican growers of the various systems of protected agriculture, they will be able to extend their growing season and their ability to supply buyers of produce over a longer period.

Imports of tomatoes from Israel are increasing each year, although only sporadically. The tonnage of tomatoes exported from Israel to Europe and the United States has more than doubled in four years - 3600 tonnes in 1985-86 to 8400 tonnes in 1989-90. The cost to air freight tomatoes from Israel to Europe is $400/tonne and $600/tonne to the United States. The Netherlands is also rapidly becoming an exporter of certain products, such as tomatoes and colored bell peppers, to the United States.

Europe. Competition in Europe is much fiercer than in North America. The Netherlands is, by far, the greatest exporter of vegetables and flowers, mainly from greenhouses.

Holland has about 7,000 greenhouse vegetable farms with 4,600 ha. under glass. About 80 percent of the production is exported; the principle export markets being Germany, Great Britain, France and Scandinavia. In 1990, Germany imported the largest amount of Dutch produce, 24 percent of the Dutch produce exports; Great Britain received 7.4 percent. The United States absorbs less than one-third of 1 percent of the Netherlands produce. Colored bell peppers, the largest volume Dutch commodity to reach the U.S. market, accounts for 8 percent of that figure.

The winter market in Europe, especially tomatoes, is mainly supplied by the Canary Islands and Spain. Through the 1960's the greenhouse area in Belgium and northern France expanded, somewhat increasing competition (Dalrymple 1973).

Other countries such as Italy, Greece, Turkey, Israel, Morocco, and Egypt all have an expanding greenhouse industry, and all look to Europe as a market for their produce. Many of these countries have historical relationships which give producers somewhat privileged access to markets. Since the early colonial days, France has had a close association with Morocco. Because of this relationship, France is a premier market for Moroccan greenhouse products. Similarly, Cyprus has close linkages to Great Britain, as does Israel to all of Europe. Such relationships do not break down easily once established over a period of years.

Competition from produce coming into Europe from southern Europe, North Africa and the Middle East will continue to grow as cultural marketing practices improve.

In Eastern Europe, during the years of communist rule, the greenhouse industry expanded rapidly, especially in Hungary, Romania, and Bulgaria. Many of the greenhouses were state-owned and controlled, and were not built to compete with the free market economy. With the collapse of communism and the opening of markets in Eastern Europe, many of these greenhouse complexes are not competitive with production in Western Europe. As a result, greenhouse horticultural products, flowers and vegetables, are flowing in from Holland.

Until the greenhouse industry in Eastern Europe is able to modernize, there will be opportunities for other nations to expand their export market in that direction. It may take a vast amount of monetary resources to rebuild the industry in Eastern Europe; however, because growers already have the horticultural expertise in place, they should be capable of recovering quickly.

A more comprehensive and detailed Annex on "Marketing and Distribution of Protected Crops," is written by Alan J. Malter, market researcher for horticultural products in the Market Research Department of Israel's Ministry of Agriculture.

13

TECHNOLOGY TRANSFER BETWEEN NATIONS

Systems of protected agriculture are used throughout the world, even if only a polyethylene roof to protect the crops from tropical rains. This chapter will concentrate on those countries in which such agricultural systems are important to the regional economy.

It is very difficult to obtain any statistics at all for many nations. Although the following discussion will not cover every nation using protected agriculture, it will give a reasonable idea of the size and nature of protected agriculture production around the globe.

Table 37. Nations serving as major sources or recipients of protected agriculture technology.

Sources of Technology	Recipients of Technology
The Americas	The Americas
Canada	Chile
United States	Colombia
Western Europe	Western Europe
England	Belgium
France	Greece
Netherlands	Portugal
Italy	
Spain	Eastern Europe
	Most countries
Eastern Europe	
Hungary	Africa and Middle East
	Algeria
	Canary Islands
Africa and Middle East	Egypt
Israel	Morocco
Turkey	Tunisia
	Iran
Asia and Oceania	Asia and Oceania
Australia	People's Republic of
Japan	China
	South Korea

Since neighboring nations often share similar climatic conditions as well as programs in protected agriculture, the countries discussed have been placed in regional groupings: the Americas, Western Europe, Eastern Europe, Africa and the Middle East, and Asia and Oceania. These regions have been further divided into nations which serve as major technology sources of information pertaining to protected agriculture and those nations which are major recipients of information.

In the following discussions, a distinction is made between major sources and major recipients of protected agriculture technology. This distinction is based on an examination of a nation's history of private and public sponsored research and its record of publications in national and international trade magazines and journals. China, for example, has vast areas devoted to protected agriculture, but does not yet have a significant history of research and publications. Therefore, China is categorized, for the purposes of this discussion, as a major recipient nation, while France, Japan, and the United States are technological sources.

Statistical data on the hectarage devoted to plastic mulch, row covers, and greenhouses for each region is detailed in Chapter II. Table 37 lists the major sources and the major recipients of this technology.

The following review of those nations listed in Table 38 will provide key examples of the use of protected agricultural systems.

THE AMERICAS

Canada and the United States are major sources of information, while Chile and Colombia are major recipients of technology. While it is difficult to obtain precise statistics for each country, Table 38 lists the approximate hectares of plastic greenhouses used in the Americas.

Table 38. Plastic greenhouses in the Americas: 1990-91

Country	ha.	Country	ha.
Canada	250	Costa Rica	200
Chile	1600	Ecuador	80
Colombia	2600	USA	4250

Source: CIPA

Canada. Canada has more than 400 ha. of greenhouses (Ingratta et.al. 1985), located primarily in British Colombia, Alberta, Nova Scotia, Ontario, and Quebec. Ontario alone accounts for more than 250 ha., with some 621 growers. Less area is devoted to vegetables. Floricultural crops predominate: In 1983, there were approximately 600 greenhouse vegetable growers producing mostly tomatoes and cucumbers valued at $45 million versus 1,000 flower growers producing a value of $225 million.

Chile. Details about the use of protected agriculture in Chile are scarce, but the steady increase in the quantity of fruits and vegetables exported to North America during the Chilean summer (winter in North America) suggests stepped-up greenhouse production. Some 1,600 ha. are under plastic greenhouses, and more than 9,000 ha. have been converted to drip irrigation for the production of vegetable and fruit crops.

Colombia. In Colombia, greenhouses are basically only rain shelters to aid in the production of floricultural crops. Colombia ranks second to the Netherlands in the export of cut flowers, accounting for 9 percent of the world market in 1989. Together, the Netherlands and Colombia dominate the U.S. market, with an 82 percent share.

Countries such as Colombia, which can grow flowers under cover without heating and cooling requirements, enjoy significant energy and investment savings. For example, a Colombian grower has minimal energy expenditures compared to a typical rose grower in the United States where, in 1988, energy costs constituted 25 percent of total expenditures. Colombia's flower production got a boost in the 1970's when the sudden increase in energy prices seriously inflated production costs in Europe and the U.S. Counterbalancing low energy costs, however, are shipping costs to North American markets.

Economies of scale are not significant, although Colombian growers believe the ideal size of a flower operation is 13 to 40 hectares, with some economies in grading and sorting. There are now more than 400 Colombian flower growers, 80 percent of whom are in the Bogota area (Morrow 1989). In 1981 the ten largest producers possessed more than 50 percent of the land devoted to floriculture, each holding from 100-150 ha. Almost 70,000 people are directly employed full-time in the industry and another 50,000 in ancillary industries such as

packaging and transportation. Flower production has provided employment to many of the poorer people in Bogota, most of whom are untrained rural women.

Colombia is well located for access to North America's large market. By air, it is only a couple of hours from Bogota to Miami. In Miami, an overland transport system was already in place when Colombia began exporting flowers to the U.S. This system transports Florida-produced flowers throughout the country: tying into this network gave Colombia complete access to the entire U.S. market.

United States. While the U.S. has a long history in protected agriculture, the industry itself is modest in size. The wide variation in growing conditions in the U.S. makes outdoor culture of food crops feasible almost any time of the year. Furthermore, the well-developed transportation system permits easy distribution of fresh produce within all regions of the United States.

Because of the relatively small scale of the protected agriculture industry, statistics on use of plastics, especially row covers, have not been systematically compiled.

Over 80,000 ha. are mulched, mostly for vegetable production: the exceptions are in California, where mulch is used for strawberry production, and in Hawaii, for pineapples. Florida has more than 40,000 ha. of mulched vegetables. In the upper midwestern states, most of the mulched crops are high value vegetables grown on family-owned farms for local roadside and city markets (Schales 1990). The principle crops mulched are listed in Table 39.

In Colombia, structures covered with polyethylene protect flower crops from rain damage.

Table 39. Principle crops mulched in the United States

Crop	Hectares
Cucumber	1,283
Eggplant	1,145
Muskmelon	6,948
Pepper	10,200
Pineapple	6,500
Specialty melons	40
Strawberries	9,234
Summer Squash	406
Sweet corn	456
Tomato	23,371
Watermelon	10,024
Other vegetable crops	15,343
TOTAL	**84,947**

Source: Schales, 1990

The trend in row covers shows increasing use of wide nonsupported or "floating" covers. These are mostly lightweight spun-bonded or non-woven fabrics in widths up to 12.8 meters, weighing as little as 8 grams per square meter. They are used to favorably modify the microclimate and to protect crops from insect vectors of crop diseases.

For greenhouse production, excellent records are kept by the United States Department of Agriculture, which document the number of growers and the major growing areas by state (Johnson 1990). Statistics for floricultural production show that in 1988, Florida and California each had more than 1,000 growers. The average greenhouse size per grower in the U.S. approximated 0.5 ha., with California growers averaging 1.1 ha. Data was not available from all 50 states: however, the major states were accounted for, totaling 3,730 ha. of greenhouses used for floricultural crops production. Of this area, 726 ha. were glass, 1,028 ha. fiberglass or other rigid materials, and 1,976 ha. of plastic film, excluding structures used for bulbs, seeds, ground covers, etc. In 1988, the total value of floricultural crops in the U.S. approximated $2.3 billion. Bedding plants totaled $755 million, followed by potted flowering plants ($508 million), foliage plants ($482 million) and cut flowers ($458 million).

Data on vegetable production in the United States shows that in 1987, 185 ha. of greenhouses were used for this purpose, mostly in California, Ohio, and Florida. The 1987 area was actually a slight decline from the 191 ha. under production in 1982.

In 1971, according to Dalrymple (1973), there were nearly 600 ha. of greenhouses devoted to vegetable production. The continuing decrease in the greenhouse vegetable industry in the U.S. is in response to the rapid increases in the cost of energy during the early 1970's

Conversely, the area devoted to greenhouse floriculture and nursery production doubled from 1967 to 1987. In 1987, 21,055 farms had 71 million square meters of area under protection, compared with 13,674 farms in 1969 with 30.5 million square meters. The greenhouse and nursery industry in the United States has indeed experienced dramatic growth, due to the higher value of floricultural crops per unit area of greenhouse, in comparison to greenhouse vegetable crops.

WESTERN EUROPE

England, France, the Netherlands, Italy, and Spain all have a long history of protected agriculture. Table 40 lists the number of hectares in nearly all those countries where plastics are used for protected agriculture.

Nearly all the greenhouses in the Netherlands are made of glass, as are many in England. Plastic predominates in France, Spain and Italy, due to their greater amounts of winter light. Much of the early research on the use of plastics in agriculture was conducted in France. Spain, while coming late to the use of plastic greenhouses, today leads Europe in the area planted to greenhouses.

Table 40. Plastic greenhouses in western Europe 1990-91 (ha.)

Country	ha.	Country	ha.
Belgium[1]	175	The Netherlands	300
Cyprus	300	Ireland[1]	50
Denmark[1]	100	Italy	18,500
England[1]	350	Norway[1]	25
Finland	1,000	Portugal	2,500
France	5,800	Spain	25,000
Germany[1]	300	Sweden[1]	50
Greece	4,240	Switzerland	50

[1]Glasshouses are either as common or more common than plastic greenhouses.
Source: CIPA

Belgium. Belgium has a relatively large protected agriculture industry. As with the Netherlands, much of the production is exported. Belgium currently has more than 2,200 ha. of glass greenhouses but is demonstrating increased interest in the less expensive plastic. Glasshouse production in Belgium has shifted from fruit to vegetable production, for the most part; 1,600 ha. are planted to vegetables and the remaining 600 ha. are planted to ornamentals.

As shown in Table 40, plastic greenhouses cover 175 ha. Approximately 500 ha. of the greenhouse vegetables are planted to soilless media, 470 ha. to rockwool and 30 ha. to NFT. Plastic greenhouses are generally used for the forcing of strawberries, followed by cucumbers, or a late strawberry crop.

Most of the mulching is with black polyethylene, covering 2,000 ha. Row covers are used on about 300 ha. and there are 2,500 ha. of floating row covers, mostly 0.05 mm polyethylene with 500 holes/m^2 for ventilation.

Exports, principally to France and Germany, consist mostly of lettuce, followed by tomatoes and then cucumbers.

England. British growers have long used glasshouses and have been slow to recognize the potential of plastics in agriculture. The 350 ha. of plastic greenhouses in England represent only about 10 percent of the total area covered.

The greenhouse industry started in the late 1800's with simple wooden structures covered with glass. In 1982, the greenhouse area was approximately 2,718 ha. Of this, 797 ha. were used to grow tomatoes, 233 ha. for cucumbers, 1,375 ha. for lettuce, and 313 ha. for other vegetables. Of this area, 24 percent of the greenhouses were built before 1960, 52 percent between 1960 and 1973, and 25 percent between 1974 and 1982 (Girard 1984). Since 1982, growth has been mainly in plastic structures with increasing usage of plastic for semi-forcing of vegetable crops and winter protection of nursery stock.

England has been a leader in adopting energy-saving techniques to offset increased fuel costs since the early 1970's. Thermal screens have been widely adopted with the use of polyester/aluminum materials. Since the first use of perforated polyethylene film in 1982, the area covered has increased annually by more than 1,000 ha. By 1990, the area using floating, or direct covers, was nearly 10,000 ha. Most of these covers are made of polyethylene, but the use of non-woven materials is increasing (Brighton 1990).

The wholesale markets still play a very significant role in the distribution of produce. These markets are sited principally in or near large towns or cities. They receive local and imported produce which is consigned to them to be sold, on commission, to secondary wholesalers and retailers. The role of these markets has been declining, however, as a greater proportion of produce is going directly from the grower to large supermarket chains.

France. Since 1965, France has been a major developer of protected agriculture technology. France's leadership, and growth, in all aspects of protected agriculture are illustrated in Table 41.

Due to the advances in climatic control under cover, the use of soilless techniques has brought more area into the cultivation of greenhouse crops.

Greece. Since joining the EEC, Greece has experienced substantial growth in the greenhouse industry. In 1970, the area devoted to plastic structures was 1,200 hectares. Twenty years later the greenhouse industry grew to over 4,200 hectares.

Glasshouses are expensive in Greece. Because of this, and the low income of the vegetable growers, nearly all of the greenhouses are of plastic. They are usually very simple in construction, consisting of a wooden frame with vertical supports or tubular curvilinear frames (high tunnels). The structures are unheated (Dalrymple 1973).

Table 41. Growth of protected agriculture in France

System	Year/hectares			
	1965	1975	1985	1988/89
		-hectares-		
Plastic mulch	2,500	35,000	80,000	100,000
Row Covers				
Supported	1,600	7,000	22,500	24,850
Floating	—	20	4,500	8,000
Greenhouses[1]	100	1,700	5,000	5,540

[1]*Plastic greenhouses - Source: Brun and Printz, 1990*

The main crops are cucumbers and tomatoes. Other crops are eggplant, peppers, squash, and strawberries.

A large area is planted to row covers. Recent data is scarce; 20 years ago, Greece had some 1,266 ha. of protected agriculture, used primarily for watermelons (Dalrymple 1973).

The Netherlands. The Netherlands has a long history in the greenhouse industry. Almost all houses are of glass; plastic mulch and row covers are rarely used.

While many growing technologies have been invented in neighboring countries, the research community and farmers of the Netherlands are experts in putting new ideas into practical use. A good example is the use of rockwool as a growing medium. Rockwool was invented in Denmark. It was perfected for use, on a large scale, in the Netherlands. The Netherlands' large research community provides valuable technology that has made the Dutch farmers some of the largest producers of vegetables and flowers, for export, in the world. The growth of the Netherlands greenhouse industry is shown in Table 42.

Early in the century, the greenhouse area was modest. After 1912, it grew rapidly, especially in vegetables and fruit production. There was little change between 1940 and 1950, due to the recovery period after the war; since then, the industry has expanded sharply in vegetables and flower production.

Table 42. Estimated utilization of greenhouse area of the Netherlands, 1912-1990[1]

Year	Total	Vegetables	Fruit	Flowers & Ornamentals
			-hectares-	
1912	194	109	85	—
1940	2367	1214	866	287
1950	2342	1270	789	283
1960	4048	3077	474	498
1965	5960	4742	316	902
1970	7249	5374	211	662
1990	9600	4800	NA[2]	4800

[1]*Mostly taken from Dalrymple (1973).*
[2]*Not available.*

Until the 1980's, much more area was devoted to vegetable production than flowers. Since then, increased competition from southern Europe and North Africa has caused many growers to switch from vegetables to flowers, especially from 1975-1980 in the Westland region of the Netherlands. The moderate climate in the Netherlands is excellent for high quality flower production. This climatic advantage over many other countries in Europe gives the Netherlands a competitive edge in floriculture which it lacks in vegetable production, which requires higher light levels. This is especially so in tomato and cucumber production.

Since 1940, the area of fruit production has steadily diminished. Grapes and tender tree fruits such as peaches and plums were once the major fruit crops. Increased competition from imported produce and obsolescence of the glasshouses used to grow the fruit has made fruit production unprofitable. It was not economical to replace the structures and plantings when cropping alternatives existed. Also, it was a problem to obtain sufficient labor for the short period involved in grape thinning (Dalrymple 1973). Fruit production in glasshouses therefore is fast becoming a thing of the past.

Of the 9,600 ha. of greenhouses, 4,800 ha. are for vegetable production (Table 43). In all there are about 7,000 greenhouse vegetable farms growing tomatoes, Boston lettuce, cucumbers, radishes and bell peppers in a dozen colors, accounting for 80 percent of the total production. The remaining 20 percent of production includes 120 vegetables, such as iceberg lettuce, bok choy, eggplant, Chinese cabbage, fennel, zucchini, broccoli, and radishes (Anon. 1992).

Rockwool was introduced in 1975, after being used on a limited scale in Scandinavia in the early 1970's. In 1988, 3,500 ha. of greenhouse crops were grown on artificial substrates; 2,500 ha. were vegetables and cut flowers and 1,000 ha. were pot plants on peat mixtures. Rockwool was used, therefore, for 40 percent of the total greenhouse area. The transition from soil to rockwool, especially with cut flowers, is expected to accelerate in the near future (Sonneveld 1988).

About 80 percent of the vegetable production is exported. Primary markets are Germany, Great Britain, France, and Scandinavia. In 1991, a total 818 million kg. of vegetables were exported; Germany received 24 percent and Great Britain 7.4 percent.

The U.S. is a growing market for Dutch produce, especially peppers. Pepper shipments have grown from 4,000 tonnes in 1986 to 10,000 tonnes in 1991. In 1991, the Central Bureau of Fruit and Vegetable Auctions in the Netherlands dedicated one-fifth of the Bureau's $1.4 million pepper promotion to developing the U.S. market.

Nearly all the growers sell through the Dutch auction system, which is successful because of a careful quality grading system. By January 1992, all the auctions were connected by a computer system. Buyers, representing many wholesalers and retailers, bid against a clock that starts at a high price. As the clock needle falls rapidly across descending prices, buyers punch a button with split-second timing to ensure they get the desired quantity at a price they are willing to pay.

The auctions constantly test products for shelf life and postharvest durability. Auction cooperatives are funded by grower fees based on a percentage of sales. Growers are also charged for specific services, such as precooling. Grower-member dues pay for all the national promotion and research projects. The growers receive payment on the Wednesday of the week following the sales of their products.

The auction bureau produces daily and weekly reports on auction volumes and prices. It also tracks harvest expectations three times a year, getting data from growers about area planted and planting date information.

Greenhouse flower production and marketing in the Netherlands is the most advanced in the world. In 1987, the Netherlands supplied 71 percent of the total world exports, up from 66 percent in 1980. In general, the Dutch have focused all their resources on a strategy to provide a constantly expanding range of top quality flowers to the world marketplace (Morrow 1989). Dutch success is attributed to a number of factors:

- A strong domestic industry with a long history of flowers production.
- An economy that is historically export-oriented.
- Year-around availability of a broad variety of flowers.
- Continual innovation as a result of extensive research and development.
- Geographical location.
- A sophisticated flower auction (Aalsmeer).
- Promotional efforts.

The Dutch floricultural industry provides considerable employment, with more than 16,000 small producers growing flowers on as little as one-tenth of a hectare of land, and many more employed in related activities. Such farmers are able to net more than $45,000 U.S. per year as a result of technologically sophisticated and intensive production practices.

The 1987 value of the export cut flower market to the Netherlands was $1.3 billion U.S. (fob). The value for Colombia, the world's second largest exporter, was $174 million U.S. Table 43 shows market shares for these and other flower exporters.

For the Netherlands to maintain its position as the preeminent flower and vegetable exporter of the world, the producers have had to be innovative and responsive to market changes. Technological improvements have allowed the Dutch growers to produce flowers previously unavailable in The Netherlands, and in other countries, and to greatly extend the growing season despite adverse climatic factors.

Continued success is also due to yearly trade fairs, advertising, horticultural exhibitions, and educational programs.

Italy. Although it is not well known, Italy has the second largest area under cover in Europe. The greenhouse industry expanded from 624 ha. in 1960 to over 5,000 ha. in 1970, and then to 18,500 ha. in 1990. Most of this tremendous growth has been under plastic. More than 50 percent of the total area is heated and most of the production is in vegetable crops.

Nearly all of the expansion in vegetable production has been absorbed internally. Imports were reduced and exports showed no particular growth (Dalrymple 1973). Italy is also a large user of plastics for mulching and row covers. Data on exact area covered by plastics is not available.

Portugal. In 1970, protected agriculture in Portugal consisted only of low plastic row covers for vegetable production and

approximately 88 ha. of greenhouses used exclusively for flowers (Dalrymple 1973). In 1990 the International Committee on Agriculture Plastics in Paris, estimated that Portugal has more than 2,500 ha. of plastic greenhouses.

Table 43. Cut flowers - market shares (%) of values of worldwide exports

Suppliers	1979	1981	1983	1985	1987
Netherlands	64	64	62	65	71
Colombia	7	10	11	12	9
Italy	11	7	8	5	5
Israel	8	8	7	6	6
Spain	1	1	2	2	4
Thailand	2	2	1	2	1
France	2	1	1	1	1
Kenya	1	1	1	1	1
USA	1	1	1	1	0
TOTAL	**95**	**95**	**92**	**93**	**97**

Source: Morrow, 1987.

Spain. The Canary Islands, off the coast of Morocco, and the S.E. Mediterranean regions were the first Spanish regions to use protected agriculture. For years, the Canaries have been a major supplier of horticulture products to Europe during the winter months. Plastics were first introduced for agricultural use in Spain in the late 1960's. Plasticulture was quickly adopted in the province of Almeria, where previous attempts to develop the region for agriculture were unsuccessful. The ground was practically uncultivated, and vegetation was scarce and often devastated by grasshoppers. While the climate is warm, water is scarce, the soil is often saline and strong winds frequently buffet the area.

Greenhouses. Almeria today has been transformed from a very poor region into immense sea of 14,500 ha. of plastic greenhouses. This region produces the earliest crops in Europe, some of which are consumed internally and the remainder exported.

While the region was nonproductive, residents often migrated to other areas of Spain. Today, 90,000 people reside in Almeria; and more than 60 banks and savings organizations operate in the area, which is an indication of new prosperity.

Lack of water for irrigation has inhibited the rate of increase in Almeria. However, in the adjacent provinces of Murcia, Alicante, and Granada, with more abundant water and a similar climate the area covered by greenhouses is rapidly increasing.

In 1970, there were only 300 ha. of plastic covered structures in Spain (Dalrymple 1973). Almost 30 years later, this area has grown to more than 25,000 ha. (Table 44).

Mulches. Polyethylene film coverage has risen dramatically as well. In 1983, mulch was used over an area of 32,800 ha.; by 1987 this figure exceeded 53,000 ha., for an annual growth rate of 16 percent. Almost all cotton and melons are grown on mulch, and other produce, such as asparagus, are also being grown on a mulched soil.

Table 44. Growth of protected agriculture in Spain, 1967 to 1990

System	1967	1971	1976	1980	1985	1990
				- hectares -		
Mulch	—	4,000	17,000	26,000	40,000	56,000
Row Covers	—	2,500	3,500	4,500	11,000	19,000
Greenhouses	—	2,000	5,000	10,500	20,000	25,000

Source: Robledo de Pedro, 1989

Table 45. Plastic greenhouses in Eastern Europe 1990-1991.

Country	ha.	Country	ha.
Bulgaria	1,350	Romania	3,500
Former Czechoslovakia	750	Former U.S.S.R.	4,800
Hungary	5,500	Yugoslavia	440
Poland	1,500		

Source: CIPA

Row Covers. From 1983 to 1988, the use of row covers increased 40 percent per year, from 6,200 ha. to 16,750 ha. This increase is due to two factors:
- The growth of strawberry cultivation and the use of low tunnels inside greenhouses for the production of crops such as early melons and watermelons, which are planted in January.
- Wide climate variations in January during the preceding three years: heavy frost and severe drops in temperatures (as low as -6°C in 1985) in warmer regions of Spain caused considerable crop damage.

Drip Irrigation. The most important use of drip irrigation in Spain is for fruit trees, with 68,000 ha. under cultivation. Citrus fruit account for 30,000 ha., and vegetables are in third place, with 27,000 ha., of which 15,000 ha. are in greenhouses. All the area inside these greenhouses is watered with drip irrigation.

There are several reasons for the rapid growth of protected agriculture in Spain. First, although Spain has warm and sunny agricultural regions, many areas receive less than 200 mm of rain per year.

Drip irrigation has proven the most efficient means of applying water in water-deficient areas. Plastics permit the production of very early cops, such as flowers, strawberries, asparagus, and other vegetables. The use of plastics is greatest along the Mediterranean.

Other factors contributing to the growth of protected agriculture in Spain are the very high rate of domestic consumption of vegetables and the close proximity of a ready market for Spain's high value winter crops.

Thanks to improved technology, less profitable crops have been replaced by more profitable ones.

Land once used entirely for cereal production is now used to grow crops of higher value, such as strawberries, melons, asparagus, and even cotton. In the southern province of Huelva, over 4,000 ha. of strawberries have replaced cereal crops.

Undoubtedly, Spain is the European leader in the application of protected agricultural systems in the production of both horticultural and agronomic crops.

EASTERN EUROPE

Since 1960, there has been a sharp expansion of greenhouse construction in the areas of the former the Soviet Union and the other nations of Eastern Europe. Table 45 lists the approximate hectares of plastic greenhouses in Eastern Europe.

Hungary. Of all the countries in Eastern Europe, Hungary may be the most active in the research and development of pro-

tected agriculture. Such research began in 1958 and Hungarian scientists have worked closely with the plastics industry ever since.

Because Hungary lacks fuels for heating greenhouses, researchers have developed a method using groundwater and thermal water as an economical way to warm greenhouses. The "Hydrosol" film greenhouse has a double wall of polyethylene. Groundwater from surrounding wells is sprayed between the film layers at the peak of the structure from a perforated pipe. The water flowing down the plastic film loses its energy between the layers of plastic, thus forming a heat insulating layer. Vegetables can be kept frost-free in these houses even in the coldest of Hungarian winters. The tomato fruit ripens is 30-40 days earlier using this system than in a single film house without heating.

In 1965, Hungary had a greenhouse area of 670 ha. By 1986 this had grown to more than 4,500 ha.; by 1990, to 5,500 ha. Only about 100 ha. are glass. The major crops grown in the plastic greenhouses are tomatoes, peppers, lettuce, spinach, parsley, and sorrel (Dalrymple 1973).

Since the mid-1970's, floating row covers have been widely adopted for lettuce, cabbage, cauliflower, kohl-rabi, and early potatoes, as well as those crops needing more warmth, such as pepper, tomato, cucumber, and melons. In 1990 it was estimated that over 1,000 ha. of floating covers were used in Hungary.

Other Eastern European Countries. Recent statistical information is not available for most of the countries. Dalrymple (1973) reported that the greenhouse industry in the former **Soviet Union** grew from 479 ha. in 1965 to 2166 ha. in 1970. Plastic structures are common, and most of the greenhouses are located in the central and northern regions.

During the 1960's, extensive greenhouse facilities were in operation on large state or collective farms close to the larger cities. The largest of these, the Moscow State Farm, has nine units, each nearly 6 ha., for a total area of 54 ha. Other large greenhouses are found outside St. Petersburg (Leningrad) (40 ha.), Kislovodsk, Simferopol, Minsk, and Kiev. Several smaller specialized farms (39) were on the outskirts of Moscow (Dalrymple, 1973). The most advanced greenhouses were purchased from the Netherlands.

In the former **Czechoslovakia**, now the Czech Republic and Slovakia, many of the greenhouse facilities are state-owned. Large greenhouse complexes were connected to electrical power plants, with warm water from the generating plant used for heating the greenhouses. The facilities were not designed for competition with a free market economy, since they were

In China, research trials are conducted on the use of protected agriculture systems.

Israel. Israel has a wide range of horticultural crops under plastic. The greenhouses range from very simple structures (high tunnels) to houses with highly sophisticated climate control equipment, which are used almost entirely for flowers. Much of this production is exported to Europe.

In 1987, Israel was the third largest cut flower exporter in the world (approximately U.S. $122 million). Given the country's favorable climate, Israeli growers were able to produce flowers at a lower price by using protective netting or plastic rather than energy intensive glasshouses (Morrow 1989).

Today, production is dependent on over 3,500 growers who work an average area of .2 to .4 hectares. Most growers live in cooperative villages, or "moshavs." These producers are served by a well-organized extension service and a marketing and distribution network co-owned by the government and the producers. To remain competitive in the world market, Israelis will have to continue to reduce the labor input and other production costs, diversify into flowers with higher returns, maintain close contact with their markets, and continue their research and development efforts to produce new and better markets.

heavily subsidized by the state. Since the reconstruction of the Czech government and the opening of the borders to imports, the Dutch have flooded the market with greenhouse vegetables and flowers, products cheaper to produce abroad than in the two republics.

The state operates 350 ha. of plastic greenhouses as well as 300 ha. of glasshouses, which are used mainly for the production of ornamentals (Anon. 1990).

The private sector will be more ready to modernize the greenhouse industry and position the industry for free market competition. The private sector already accounts for more than half of the area of plastic greenhouses — the 400 ha. located mainly in south Slovakia.

Floating cover (35 ha.) is also expected to help with the development of protected agriculture in a free market.

Recent agricultural advances in **Bulgaria,** especially in horticulture and tobacco production, have been closely linked with the increased use of protected agriculture. Low density polyethylene and PVC are the materials most widely used for covering crops (greenhouses, low tunnels, and floating covers), for mulch, and thermal screens. In 1986, the total area for protected agriculture was 2,250 ha. Plastic greenhouses account for some 1,300 ha. and produce more than 130,000 tonnes of fresh vegetables, i.e., 14-15 kg. per inhabitant. In addition to plastic greenhouses, there are 50 ha. of glasshouses (Tzekleev et.al. 1988). The application of plastics is rapidly extending into flower production, fruit tree nurseries, viticulture, forestry, and other crops.

Current information about **Poland, Romania**, and **Yugoslavia** is not readily available, except for those statistics listed in Table 45.

AFRICA AND THE MIDDLE EAST
Israel is the only country in this region which has had an extensive research and development program in protected agriculture. Table 45 lists the approximate hectares of plastic greenhouses in Africa and the Middle East.

Table 46. Plastic greenhouses in Africa and the Middle East, 1990-1991

Country	ha.	Country	ha.
Algeria	5,000	Libya	1,000
Egypt	1,000	Morocco	3,000
Israel	1,500	South Africa	200
Jordan	450	Tunisia	1,100
		Turkey	9,800

Source: CIPA

Turkey. While it is not widely recognized, Turkey has a large and expanding greenhouse industry. Turkey constructed its first greenhouse at the coastal city of Antalya in 1940 (Dalrymple 1973). In 1965, there were 600 hectares of protected agriculture, by 1972, 2,000 ha.; and by 1990, more than 9,800 ha.

Since 1968, Turkey has received substantial assistance from the United Nations Development Program, with emphasis on determining the most applicable systems of greenhouse production for local conditions. This investigation led to improved methods of construction, heating, and watering.

Middle East Countries. Algeria, Morocco, Tunisia and Egypt have all demonstrated great increases in the use of protected agriculture. The only statistical data available are the estimates in Table 46.

ASIA AND OCEANIA

Asia is by far the largest user of protected agriculture. In China, whether in a rural or urban setting, one sees thousands of hectares covered with plastic as mulch, row covers, and greenhouses.

Table 47 lists the hectares of the different systems used in each of the major users of protected agriculture in the regions of Asia and Oceania.

Australia. Australia's protected agricultural industry is small, primarily because of the country's low population. Nevertheless, Australia has had an extensive research program in greenhouse agriculture, mainly through the government-operated CSIRO. National data on protected agriculture is not available.

China. Approximately 2,867,000 hectares of plastic mulch are used in China on all types of vegetable crops, as well as cotton, sugarcane, corn, and peanuts. China has more than 80,000 ha. of plastic row covers and 43,000 ha. of plastic greenhouses.

Cucumber cultivars, specifically bred for greenhouse production, are grown mostly in soil.

Table 47. Protected agriculture in Asia and Oceania, 1990-1991

Country	Mulch	Row Covers	Greenhouses
		-hectares-	
Australia	NA[1]	NA	600
Japan	195,000	61,000	42,000
China (PRC)	2,867,000	80,000	48,000
South Korea	267,585	9,500	22,000

[1]*Not available.*

Various hydroponic methods, such as the NFT system, are being tested in China for greenhouse vegetable production.

Another 5,000 ha. are covered by traditional Chinese lean-to-wall type greenhouses, used for the most part to grow vegetable seedlings. However, only 50 ha. of large modern glass greenhouses presently exist in China, mainly in the northeastern section of the country. With this exception, all greenhouses are unheated except for some of the lean-to wall structures used for transplant production. Bamboo is commonly used for greenhouses in most areas of China, since money for steel pipe is limited.

Plastics are used primarily to extend the growing season, both in the spring and the fall. In many areas of China, 80 percent of the farmers use plastic mulch; some farmers have over 70 percent of their crops planted this way. In the spring, up to 30 percent of the vegetable crops are covered with plastic row covers. Many vegetable crops are commonly started in plastic greenhouses; the plastic is removed in the early summer to facilitate ventilation. Plastics for greenhouse use are either polyethylene or vinyl. The mulch is normally made of polyethylene in a variety of colors, with clear predominating. The Chinese government subsidizes the cost of the plastics to the farmers.

The use of plastic row covers for vegetable production began in 1972 and plastic mulch was introduced in 1978. This innovation spread quickly in the next couple of years (Table 48). The rapid growth of the greenhouse industry has made it difficult to maintain accurate, up-to-date statistics; however, in some regions, the use of plastics for greenhouses nearly doubles annually.

Table 48. The development of plastic mulch in China

Year	Area (hectares)
1979	44
1980	1,666
1981	20,920
1983	629,133
1985	1,500,000
1987	2,266,666
1989	2,867,000

Source: Huang, 1989

The government of China has an active, on-going research and development program in protected agriculture, not only on the federal scale, but also at the provincial and municipal level.

Japan. In Japan, the first methods of protected cultivation were developed in the 1600's when oil paper and straw mats were used to protect crops from the severe climatic conditions. After the Meiji era, approximately 100 years ago, the first European type glasshouse was imported and built àt the agricultural experiment station in Tokyo. The commercial use of glasshouses was limited to particular crops in areas near large cities.

It was not until 1951, after the introduction of PVC film, that protected cultivation became widespread and its benefits to agriculture fully appreciated. Paper covered tunnels were rapidly replaced by tunnels of PVC film. Traditional wooden or bamboo frames were first replaced by steel, then aluminum, and in some cases, plastic.

Today the area of Japanese greenhouse cultivation exceeds 42,000 ha. The energy crises in the early 1970's severely affected the growth rates of greenhouse vegetable production; however, this was offset by the rapid increase in pomiculture (fruit crops).

In 1976, the area of pomiculture surpassed floriculture. This growth is expected to continue, eventually to more than twice that area devoted to floricultural crops. Today, the total greenhouse area in fruit crops is 7,500 ha. This includes 1.9 percent of the total area of pomiculture as well as the hectarage planted outdoors. Viticulture accounts for 84 percent of the total area in greenhouse fruit production: oranges and pears account for much of the rest.

This rapid increase in greenhouse fruit production is because of: 1) increased yields, 2) better quality fruit, and 3) blueprint planting and harvesting. Yields and fruit quality, in particular, are dramatically improved because of better control over diseases, insects, rain,and wind.

More than 60 percent of Japanese greenhouses are unheated; they depend on the large heat mass of soil and thermal insulation properties of various films. Takakura (1988) describes the very high emissivity of PVC film for long wave radiation (similar to glass), which creates slightly higher night air temperatures in the greenhouse. This improvement in the thermal environment gives PVC films a competitive advantage over the lower priced polyethlene.

In Japan, the big advantage of plastic greenhouses is that the film can be removed during the warm summer months. With plastic, also, thermal insulation can be improved by multi-layer coverings.

The inside row covers are primarily used for growing seedlings; they are usually removed when the crop matures. Opaque sheets can also be applied at night. Floating mulches are becoming a popular alternative to inside row covers. According to Takakura (1988), more than 90 percent of the heated greenhouses have at least one layer of thermal screen which is movable.

Temporary protective structures, called "rain shelter greenhouses" in Japan have rapidly become as popular in Japan as they are in Korea. In Japan, they have found special application in fruit and leafy vegetable production. Of the 2,400 ha. covered this way in 1983, 85 percent was planted in tomatoes and spinach. These structures have drastically reduced the incidence of disease and fruit injury, especially bacterial canker on tomato and downy mildew on spinach.

Greenhouse agriculture covers less than 1 percent of the total arable land in Japan. Despite this low percentage, greenhouses provide most of the main vegetables consumed in Japan, with significant production of tomatoes, cucumbers, green peppers, and strawberries (Table 49).

Table 49. The percent of vegetable production in greenhouses to total production (1984)

Crop	1975	1979	1983
Eggplant	32%	35%	37%
Tomatoes	34	40	54
Cucumbers	49	53	57
Pumpkins	31	37	37
Green Peppers	60	66	65
Strawberries	81	85	89
Watermelons	63	73	77
Lettuce	28	30	32

Source: Takakura, 1988

South Korea. In Korea, the cultivation of crops in protected structures began in the 1920's with the use of oil paper and accelerated with the introduction of PVC film in 1952 and polyethylene in 1955 (Anon. 1983). In the last ten years, crop production in plastic covered greenhouses has multiplied six times over, from 3100 ha. in 1975 to 21,000 ha. in 1986. The area covered with plastic row covers totals 9,500 ha., an increase of 290 percent over the acreage in 1975 (Park 1988). Most plastic covered greenhouses are constructed with steel or plastic pipes that are fabricated into single- or multi-span type structures.

Crops are mostly vegetables — tomatoes, cucumbers, melons, and cabbage. A variety of ornamental crops are produced, especially cacti and orchids, as well as several temperate and tropical fruit crops.

Plastic film is particularly valuable for temporary protection in fruit production. These structures provide protection against adverse weather, including low temperatures, rain, and strong winds. They also prevent disease and damage by insects and birds, thus improving fruit quality. Korean plastic greenhouses are widely used for such crops as grapes, oranges, and pears. To force grapes for early maturity, the structural frames are installed in November, before the arrival of cold temperatures. The plastic films are added in late February or early March. Normally a system of double glazing is used. When the inside temperatures reach 30°C, the houses are ventilated by opening the covered films. The plastic film is removed in late May or early June, in order to allow more sunlight to reach the crops.

This procedure has resulted in earlier harvest and better fruit quality.

An alternate method of top covering for viticulture is to build a plastic quonset-type structure over the grape vines. This method is particularly useful during the budding and flowering stages to give temporary protection from heavy rainfall and wind.

In South Korea, vegetable production in plastic tunnels has increased six fold from 1975-1986.

14

DEVELOPMENT CONSTRAINTS, RESEARCH NEEDS AND THE FUTURE OF PROTECTED AGRICULTURE

In the next decade, there will be many new developments in protected agriculture. We can look forward to fascinating new technologies in mulching and covering materials, as well as new energy conservation methods and ways to control environments through the use of computers, artificial intelligence, and robotics. Genetic engineering and biotechnology also promise exciting developments.

Key focal points for research and development will continued to be pest and disease control, better product nutrition and appearance, improved shelf life, and new methods of harvest, rapid transportation, and marketing.

A country's competitiveness — despite increases in labor, production, transportation and marketing costs — will depend greatly on its willingness to invest in research and extension, and to search for new, innovative ideas and technologies.

How communities and countries not engaged in any system of protected agriculture might enter the market place is indeed a major challenge which deserves utmost analysis and attention.

This chapter deals with many of the development constraints, the research needs, and the future of protected agriculture.

DEVELOPMENT CONSTRAINTS

In many regions of the world, it is difficult for any system of protected agriculture to compete with open field agriculture (OFA). Economically, for example, greenhouse vegetable production in the United States is almost prohibitive for five separate but interdependent reasons:

1. The diverse climate in North America (unlike the fairly homogeneous climates of Japan and Europe) permits conventional field production of vegetables somewhere on the continent during any time of year at relatively low cost.
2. A rapid and effective nationwide transportation system makes local food production for individual regions or communities unnecessary.
3. The already high capital and energy costs for greenhouse agriculture has increased alarmingly since the mid-1970's, with no concomitant breakthroughs in CEA design or materials to mitigate these costs.
4. Few chemicals for disease and pest control have been cleared by the U.S. government for use in greenhouses, and the market is too small for manufacturers to undertake the expense of registration.
5. The technical and economic perils of greenhouses require an unusual management mixture of biological and engineering sciences, as well as informed and aggressive marketing. Such combinations are rare.

The impact of the climate is significant, and the least possible

to manipulate. Were no other vegetables readily available in the United States at low prices during domestic "off seasons," the higher costs of greenhouse vegetable production would not be such a competitive disadvantage. Parts of the southern United States and northern Mexico can and do grow winter vegetables. During the past 10 years, Mexico greatly increased OFA off-season exports to the United States, particularly of tomatoes, one of the most profitable greenhouse crops. These Mexican imports have reduced American off-season peak prices, leading to even smaller profitability margins.

Probably the most important aspect of the American transportation system, for fresh produce, is the interstate highways and related networks. Low-cost, rapid, frequent, precisely-timed delivery of open-field vegetables from any growing area to any market region anywhere in the United States is now taken for granted. Since no community must provide its own vegetables, the potential for off-season, local greenhouse production is greatly blunted.

The third reason for the static greenhouse vegetable industry in the United States - the high cost of capital structures and energy inputs - has remained technically intractable because of minimal market demand. Few answers have been found because few have been sought. In the last full-scale technology assessment of CEA (Anon. 1977), a theoretical new design was postulated to reduce capital costs 60-80 percent, while reducing inputs of purchased energy to nearly zero by utilizing solar energy. No such structure has been built. The most significant technical improvement required, a longer-lived, lower-cost, selectively-transparent plastic film, is believed to be within the capability of manufacturers. No such film has been made available.

The fourth reason requires no further explanation. While European greenhouse operators use a variety of chemical controls for plant diseases and pests, and while a large number of pesticides and herbicides are cleared for OFA use in the United States, few chemicals have been approved for American greenhouses. Chemical manufacturers are unwilling to spend the large sums necessary to obtain clearance as long as the market is small.

The fifth reason for a dormant U.S. greenhouse vegetable industry - the lack of interdisciplinary support and management - is not merely a contemporary "hard times" phenomenon. This is evident by a review of hydroponic CEA commercial failures during the brief "golden" era (if it could be called that) of American hydroponics during the late 1960's. At that time, market demand was growing, purchased energy was cheap, structural materials were inexpensive, competitive field crops were not extensively imported, and high-risk investment funds were readily available. Even then, the failure rate was high.

Hydroponics is an inherently attractive, often oversimplified technology, which is far easier to promote than to sustain. A weakness in any of a number of technical or economic links snaps a complex chain. And weaknesses due to management inexperience or lack of scientific and engineering support have been common. The list of problems plaguing hydroponic operations is long: low yields, nutrient-deficient and unattractive crops, plant diseases, insect infestation, summer overheating, winter chilling, undercapitalization, odd promotion schemes, indifferent cost accounting and, the most lethal, ignorance of the subtleties of produce marketing. The energy crunches and embargoes of the early 1970's were less a mortal wound to many hydroponics operators than merely a *coup de grace.*

Given this background, one of the most surprising things about U.S. greenhouse vegetable hydroponics is that it exists. There are still strong levels of popular and corporate support. During the past five years, several American corporations have developed new prototype commercial hydroponic systems. These do not appear to have been economic successes, yet at least three other large corporations, undaunted, are known to be planning projects of their own.

Some of the same constraints may exist for growers in less developed countries (LDC), as well as a number of additional issues which may hinder many nations from entering the market place.

Many of the common problems outlined by Morrow (1989) for LDC growers in entering the market place include:
- poor reputation,
- inadequate market information, and
- lack of market presence.

A common problem is inadequate technology on the particular system of protected agriculture and the lack of management and business skills. Many countries do not have trained research and extension personnel in protected agriculture.

Poor reputation is most often based on problems of quality and reliability, including unreliable grading standards and consistency in following the even good standards. Harvesting and packing procedures including containers, refrigeration, and adequate transport vehicles and roads may also be deficient.

Experience has shown that even if production, harvesting, and packaging are letter perfect, all might be lost in the transport of the commodity to market. For example, air freight deliveries are often inadequate, indirect, and unreliable. Designated freight companies may lack experience in handling of perishables; cargo space from an LDC may be limited, or government-controlled shipping rates may make certain perishable products less desirable cargo when compared to traditional or familiar commodities. In addition, gaps in communication and information - on market prices, supplies, or trends - commonly result in missed market opportunities.

Lack of market presence may be a major problem, since buyers may be unfamiliar with the country's products and particular exporters. In the marketing system, personal acquaintanceships and reputation are very important. Unless a country has a representative in the market,it can be extremely difficult to find buyers. Countries must set up their own import offices in given countries, making direct contact with retailers in order to expand sales and establish a reputation for quality and reliability.

Finally, it is important for countries to have good ongoing research and development programs, with a mechanism to reach the farmers with new and recommended technology. This can be accomplished through a good extension service, trade fairs, and exhibits.

RESEARCH NEEDS

As plastic consumption has increased dramatically in recent years, waste plastic has become one of the biggest problems in protected agriculture. In China, large waste sites accumulate used plastics. Land fills are rapidly becoming full, and few economic alternatives exist to recycle the plastic. Incinerating large amounts of plastic without producing air and water pollution is difficult and not recommended. Incentives to recycle plastics are needed along with viable ideas for the reuse of plastics.

Photodegradable mulch films are being used increasingly, but the buried edges, which are not exposed to light, do not degrade. The effects of residues of photodegradable films after years of use on cultivated ground must be studied.

There is a definite need for the development of biodegradable films which will be technically and economically competitive. Photobiodegradation seems to be an answer for mulch film, because recovery and/or recycling methods are underdeveloped and very expensive.

Recently, a group of botanical researchers reported the development of a genetically engineered plant to produce plastic resin. This is the newest entry in a world-wide competition to develop biodegradable plastics. Researchers estimate 10-15 years of further development before we see such plastics in the fields in commercial crops.

To reduce the cost of plastic mulching, thinner films are needed that are consistent in thickness and strength. Researchers must continue to study the influence of the mulch color on crop growth and reproduction, and to develop mulches which are wavelength-selective, such as those which transmit the infrared radiation and not the visible light.

In spite of the constraints in the use of protected agriculture, the future is promising. For countries not yet using any system of protected agriculture, the future does not require dramatic scientific breakthroughs but rather a series of relatively modest technological improvements in comparative economics.

For greenhouse crops, research should concentrate on, but not be limited to, the following:

1. Design engineering of a lowest-cost CEA structure to be ventilated, heated, and cooled as much as possible by solar and solar-effect phenomena, and other alternative energy sources.
2. Materials engineering, particularly in the development of a low-cost, long-lived, selective transparent film for CEA roof structures.
3. Development of lower-cost nighttime insulation devices and techniques.
4. Design of a plant bioengineering program to develop new, temperature-tolerant, machine-harvestable, disease-resistant greenhouse cultivars.
5. Root temperature studies to determine influences on growth rates and plant development.
6. Disease control of waterborne pathogens in closed hydro-

ponic systems (filtration, UV radiation, etc.).

7. Integrated pest management systems for greenhouse applications in order to minimize the need for pesticides.

8. The development and governmental approval of chemicals for disease and pest control in greenhouses.

9. New aggregate material(s) (e.g., a counterpart of European rockwool) for lower-cost installation and maintenance.

10. Wider utilization of industrial waste heat for greenhouse heating.

While research is being conducted on most of these goals, such work is very limited and underfunded. To the extent that these improvements increase crop yield and quality and reduce unit costs of production, protected agriculture will become more competitive.

As consumers become increasingly aware of quality differences, the demand for products of protected agriculture will increase providing they have the buying power to purchase such commodities.

THE FUTURE OF PROTECTED AGRICULTURE

There seems to be a kind of technological imperative driving development of protected agriculture. Like manufacturing, it generally moves toward higher-technology, more capital-intensive solutions to problems. Protected agriculture is highly productive, suitable for automation, conservative of water and land, protective of the environment and yet, for most employees, requires only basic agricultural skills. It can be argued (and has been) that protected agriculture is "the next logical step" after traditional OFA.

Given present circumstances, however, there seems to be no rational basis for anticipating a much wider and faster diffusion of technology than is presently occurring. The future growth of protected agriculture depends greatly on the development of systems of production that are cost-competitive with those of open field agriculture.

Continuing research and development, for example, may lead to more cost-efficient structures and materials; to reduced requirements of purchased energy; to new cultivars more appropriate to controlled environments and mechanized systems; to better control (including improved plant resistance) of diseases and pests. To the extent that these improvements increase crop yield and reduce unit costs of production, protected agriculture will become more competitive.

The economic prospects for protected agriculture may change if governmental bodies determine that, in some circumstances, politically desirable effects of protected agriculture merit subsidy for the public good.

Such beneficial effects may include the conservation of water in regions of scarcity or food production in hostile environments; governmental support for these reasons has occurred in the Middle East. Another desirable societal effect can be the provision of income-producing employment for chronically disadvantaged segments of the population entrapped in economically depressed regions; such employment produces tax revenues as well as personal incomes, reducing the impact on welfare rolls and improving the quality of life.

Protected agriculture is a technical reality. Such production systems are extending the growing seasons in many regions of the world and producing horticultural crops where field-grown fresh vegetables and ornamentals are unavailable for much of the year. The economic well-being of many communities throughout the world has been enhanced by the development and use of protected agriculture. Such systems offer many new alternatives and opportunities for tomorrow's population, new systems that encourage conservation and preservation of the environment rather than the exploitation of the land and water.

REFERENCES

Aldrich, Robert A. and J. W. Bartok, Jr. 1986. Greenhouse engineering. Dept. of Agric. Eng., Univ. of Connecticut, Storrs.

Anon. 1971. Low tunnel crop protection. Grower Bulletin No. 3, British Visqueen Limited. p. 1-8.

Anon. 1977. An assessment of controlled environment agriculture technology. NSF Contract C-1026. Intern. Research & Technology Corp., McLean, Va.

Anon. 1987. LDPE Agricultural films - worldwide plasticulture. CIPA, 65, rue de Prony, Paris, France.

Anon. 1990. Making the most of available space. The Packer, Ma. 28, p. 5E-6E.

Anon. 1990. Czechoslovakia - Vegetable growers take up plastics. Plasticulture No. 85.

Anon. 1983. Plastics for agricultural uses. Korea Pacific Chem. Corp. Plastic News, 70.

Axlund, D. S., S. T. Besemer and A. Brown, 1974. Greenhouse insulation experiment. San Diego Gas and Elec. Co.

Axtmann, R. C. and L. B. Peck. 1976. Geothermal chemical engineering. Amer. Inst. Chem. Eng. J. 22:817-828.

Badger, P. C. and H. A. Poole. 1979. Conserving energy in Ohio Greenhouses. The Ohio State Univ. p. 42.

Badger, P. C. and H. A. Poole. 1979. Conserving energy in Ohio greenhouses. The Ohio State Univ. p. 41

Bonsignore, P. V., R. D. Coleman, W. W. Schertz, T. S. Tsai, and S. P. Tsai. 1990. Potato peels to degradable plastics. Proc. Nat. Agric. Plas. Cong. 22:230-235.

Bravenboer, L., Ed. 1974. Integrated control in greenhouses. Organisation Internationale de Lutte Biologique/S.R.O.P., Antibes, France

Brighton, C. A. 1990. Plastics in agriculture and horticulture in the U.K. Proc. XI Intn. Plastics Cong. 11: J49-J52.

Brun and Printz, 1990. A sector which is ever-developing in France. Proc. XI Intn. Plastics Cong. 11:J16-J19.

Buclon, Francis. 1966. Comparisons of agricultural uses of plastics in France, Italy, Japan, Russia and the United States. Proc. Nat. Agric. Plastics Cong. 7:21-33.

Burgess, H.D. 1974. Modern pest control in glasshouses. Span 17:32-36.

Carolus, R. L. 1962. Mico-climatic influence of polyethylene mulching on behaviour and productivity of warm season crops. Proc. l6th Int. Hort. Cong., Brussels. Vol. 1, p. 137.

CAST. 1975. Energy conservation in agriculture. CAST Report 40.

Change, Jen-hu. 1970. Potential photosynthesis and crop productivity. Annuals of the Assoc. of Am. Geographers. p.96.

Cornwell, J. T. 1989. The recycling of plastics in agriculture. Proc. Nat. Ag. Plastic Cong. 21:60-64.

Cotter, D. J. and C. E. Chaplin. 1967. A review of plastic greenhouses: The problems, progress, and possibilities. HortScience 2(1): 7-9.

Dalyrymple, D. G. 1973. A global review of greenhouse food production. USDA Rpt. 89.

Decoteau, D.R., D.D. Daniels, M.J. Kasperbauer, and P. G. Hunt. 1986. Colored plastic mulches and tomato morphogenesis. Proc. Nat. Agric. Plastics Cong. 19:240-246.

DuBois, P. 1978. Plastics in agriculture. Applied Sci. Pub., London.

Evans, S. G. 1979. Susceptibility of plants to fungal pathogens when grown by the nutrient film technique (NFT). Plant Path. 28(11):45-48.

Ettinger, M. J. 1964. The use of plastic tunnels and plastic greenhouses in Israel. Comite des Plastiques en Agriculture, 7th Colloquim, p. 31.

French, N., W. J. Parr, H.J. Gould, J. J. Williams, and S. P. Simmonds. 1976. Development of biological methods for the control of *Tetranychus urticae* on tomatoes using *Phytoseiulus persimilis*. Ann. Apl. Biol. 83:177-189.

Garrison, S. A. 1973. A plastic-covered trench system for increasing the earliness of vegetables. Proc. Nat. Agric. Plastics Cong. 11:1-12.

Garrison, S. A. l990. Managing degradable mulches. Amer. Veg. Grower. 38(3): 73,76.

Gent, P.N.M. 1989. Row covers to produce red or yellow peppers. The Conn. Agric. Exp. Station, New Haven, Bul. 870. p. 13.

Geraldson, C.M. 1962. Growing tomatoes and cucumber with high analysis fertilizer and plastic mulch. Proc. Fla. Hort. Soc. 75:253-260.

Giacomelli, G.A., W. J. Roberts, D. R. Mears, and H. W. Janes. 1984. Greenhouse tomato production in a movable cable supported system. Acta Hort. 148:89-95.

Gibault, George. 1912. Historie des Legumes. Librarie Horticole, Paris, p. 367-368.

Girard, A., J. 1984. An overview of the United Kingdom greenhouse vegetable industry. Proc. Amer. Greenhouse Vegetable Growers Conf. Amer. Greenhouse Veg. Growers Assoc. p. 29-38.

Glenn, E.P. 1984. Seasonal effects of radiation and temperatures on growth of greenhouse lettuce in a high isolation desert environment. Scientific Horticulture. 22:9-21.

Gould, H. J., W. J. Parr, H. C. Woodville and S. P. Simmonds. 1975. Biological control of glasshouse whitefly (*Trialeurodes vaporariorum*) on cucumbers. Entomophaga 20: 255-292.

Graves, C.J. 1983. The nutrient film technique. Horticultural Review. 5: 1-44.

Hall, B. J. 1963. Continuous polyethylene tube covers for cucumbers. Proc.Nat. Ag. Plastics Cong. 4:112-132.

Hall, B.J. 1971. Comparision of drip and furrow irrigation for market tomatoes. Proc. Nat. Ag. Plastics Cong. 10:19-27.

Hall, B. J. and S. T. Besemer. 1972. Agricultural plastics in Calif. HortScience 7(4): 373-378.

Hanger, B. 1982b. Rockwool in horticulture-a review. Austral. Hort. 80(5):7-16.

Hemphill, D.D., Jr., G. L. Reed and O. Guthrod. l987. Floating row covers prevent virus transmission in potato seed stock. Proc. Nat. Agr. Plastics Congress 20:117-121.

REFERENCES

Hibbard, R. P. 1926. Frost protectors for early planting. Mich. Quart. Bul. 7(4):150-153.

Hoag, P. 1988. Wide row covers prove cost effective. The Grower 21(1): 38-39.

Hoagland, D. R. and D. I. Arnon. 1950. The water-culture method for growing plants without soil. Cir. 347. Cal. Agric. Exp. Station, Uni. of Calif., Berkeley.

Hopen, J. H. and N. F. Oebker. 1976. Vegetable Crop Responses to Synthetic Mulches. Univ. of Illinois, Spec. Publ. 42.

Huang, Zhenbu. 1989. The research and application of plastic films in China. Chinese Plastic Mulch Research Assoc.

Ingratta, F.J., T. J. Blom and W. A. Straver, 1985. Canada: Current research and developments. Proc. Hydroponic Worldwide, Intn. Center for Special Studies, Honolulu, HI p. 95-102.

Islam, N. 1990. Horticultural exports of developing countries: past performances, future prospects, and policy issues. Intn. Food Policy Res. Inst. Wash, D.C.

Jensen, M. H. and R. Sheldrake. 1965. Concluding results of air supported row covers for early vegetable production. Proc. Nat. Agric. Plastics Cong. 6:100-112.

Jensen, M. H. and R. Sheldrake. 1967. Air supported row covers for early vegetable production. Proc. XVII Intn. Hort. Congress 111:369-378.

Jensen, M. H. and H.M. Eisa. 1972. Controlled Environment Vegetable Production: Results of Trials at Puerto Penasco, Mexico. Environmental Research Lab. Univ. of Arizona, Tucson. 117p.

Jensen, M.H. 1973. Exciting future for sand culture. Amer.Veg. Grower 21(11); 33-34, 72.

Jensen, M. H. 1975. Arizona research in controlled environment agriculture, p. 13-82,In: Proc. Tenn. Valley Greenhouse Veg. Workshop, Chattanooga, Tenn., USA.

Jensen, Merle H. 1977. Five years of intensive vegetable production on a desert seacoast. Proc. Intn. Agric. Plastics Cong. 7:24-32.

Jensen, M.H. 1982. Review of greenhouse leafy vegetable industry in Holland, England, Norway, and Denmark, Env. Res. Lab., Uni. of Arizona, Tucson.

Jensen, M.H. and W. L. Collins. 1985. Hydroponic vegetable production. Horticultural Review 7. 483-558.

Johnson, D.C. 1990. Floriculture and environmental horticulture products. USDA, Bul. No. 817.

Kao, T.C. 1990. Development of a hydroponic vegetable factory system. p. 387-390. In Proc. of 1990 Chinese American Academic and Professional Convention, Chinese Amer. Academic and Prof. Soc. N.Y., N.Y.

Kaplan, J.K. 1991. Dress-for-success mulch. Agric. Research, USDA-ARS 39(9):10-13.

Katan, J. 1981. Solar heating (solarization) of soil for control of soilborne pests. Annu. Rev. Phytopath. 19:211-236.

Kyte, L. 1987. Plants from test tubes. Timber Press, Portland, Oregon. p. 160.

Lamont, W. J. 1991. The use of plastic mulches for vegetable production. Dept. of Hort., Kansas St. Univ. p. 13.

Lauder,K. 1977. Lettuce on concrete. Grower (87)8:40-41, 44-46.

Lawson, G. 1982. New air filled pack extend shelf life to three weeks. Grower. 97(26):47-48.

LeMaire, M. 1964. Semi-forcing under perforated tunnels. Comite des Plastiques en Agriculture. 7th Colloquium pp. 9-14.

Liu, R.C. and G.E. Carlson. 1976. Proposed solar greenhouse design. Proc. Solar Energy Fuel-Food Workshop. Univ. of Arizona, Tucson. p. 129-141.

Loy, J.B. and O.S. Wells. 1974. Response of hybrid muskmelons to polyethylene row covers and black polyethylene mulch. Scientia Hort 3:223-330.

Loy, J. B. and O. S. Wells. 1982. A comparison of slitted polyethylene and spunbonded polyester for plant row covers. HortScience l7(3): 405-407.

Loy, B., J. Lindstrom, S. Gordon, D. Rudd, and O. Wells. 1989. Theory and development of wavelength selective mulches. Proc. Nat. Agric. Plastics Cong. 21:193-197.

Loy, J. B. and O. S. Wells. 1989. IRT mulch: high-tech at ground level. Amer. Veg. Grower 37(11):l4, 16-17.

Madewell, C. W., L. D. King, J. Carter, J. B. Martin, and W. K. Furlong. 1975. Using power plant discharge water in greenhouse vegetable production. Progress Rpt. Bul. Z-56, Tenn. Valley Authority.

Mansour, N. S. 1991. The use of field covers in vegetable production. Proc. of Intn. Workshop on Imp. Veg. Prod. Through the Use of Fert., Mulching and Irrigation, Chiang Mai Univ., Thailand.

Massantini, F. 1976. Floating Hydroponics; A New Method of Soilless Culture. 91-98. In: Proc. Intern Working Group on Soilless Culture, 4th Intern. Congress on Soilless Culture, Las Palmas, Canary Islands, Spain.

Mayberry, K.S. 1988. Plastic covered trenches for cantaloupes. Calif. Cooperative Ext. Service, El Centrol, Calif. p. 2.

Mears, D. R. and C. D. Baird. 1976. Development of a low-cost solar heating system for greenhouses. Proc. Solar Energy Fuel-Food Workshop, Uni. of Arizona, Tucson. p. 88-109.

Mermelstein, N.H. 1980. Innovative packaging of produce earns 1980 IFT food technology industrial achievement award. Food Technol. 34(16):42-48.

Milner, H. G. 1963. Petroleum Mulch. Nature. 197(4864): 241-242.

Morgan, J.V. and A. Tan. 1982. Production of greenhouse lettuce at high density in hydroponics. p. 1620 (Abstr.). In: Proc. 21st Intern. Hort. Cong., Hamburg, Vol. 1.

Morrow, F. 1989. Flowers: Global Subsector Study. The World Bank Industry and Energy Dept. Paper No. 17.

Nettles, V. F. 1963. Planting and mulching studies with cucurbit. Proc. Fla. St. Hort. Soc. 76:178-182.

Otey, F. H. and R. P. Westoff. 1980. Biodegradable starch-based plastic films for agricultural applications. Proc. Agric. Plast. Cong. 15:90-93.

Park. Young D. 1988. The use of plastic films in agric. Proc. Intn. Sem. on the Util. of Plastics in Agric., Rural Dev. Admin. Suweon, Republic of Korea, p. 3.1-3.15.

Peck, J. F. 1976. Basic solar collector design and considerations. Proc. Solar Energy Fuel-Food Workshop. Univ. of Arizona, Tucson, p. 15-27.

Potter, R.F. and T. Sims. 1975. New developments in chrysanthemum culture in nutrient film. Nat'l. Chrysanthemum Soc. Bull. 87:13-15.

Price, D. R., G. E. Wilson, D. P. Froehlich, and R. W. Crump. 1976. Solar heating of greenhouses in the Northeast. Proc. Solar Energy Fuel-Food Workshop. Univ. of Az., Tucson, p. 173-190.

Prince, R. P.; W. Giger, Jr.; J. W. Bartock, Jr.; and T. L. Logee. 1976. Controlled environment plant growth. Dept. Agric. Eng. Univ. of Conn., Storrs.

Roberts, W. J. and D. R. Mears. 1969. Double covering a film greenhouse using air to separate film layers. Trans. Amer. Soc. Agr. Eng. 12:32-33,38.

Roberts, W. J.,J. C. Simpkins and P. Kendall. 1976. Using solar energy to heat plastic film greenhouses. Proc. Solar Energy Fuel-Food Workshop. Uni. of Arizona, Tucson. p. 142-159.

Robledo de Pedro, F. 1989. Spanish plasticulture: soncsumption and applications of plastics. Plasticulture 82: 13-22.

Roche, H.W. 1964. Nudging nature for early profits. Amer. Veg. Grower. 12(4): 9, 10, 11, 50.

Rogers, H.T. 1983. Greenhouse of the future. Greenhouse Grower 1(1): 72, 74, 76.

Schales, F. C. and P. H. Massey. 1965. Starting early plants. Circ. 764. Ag. Ext. Sn., Virginia Poly. Inst., Blacksburg, Virginia, p. 15.

Schales, F.D. 1990. Agricultural plastics in the United States. Proc. XI Intn. Plastic Cong. 11:354-356.

Schippers, P.A. 1978. A vertical hydroponic system. Amer. Veg. Grower 26(5):20-21.

Schippers, P.A. 1982. Developments in hydroponic tomato growing. Amer. Veg. Grower 30(11):26-27.

Shakesshaft, R.G. 1981. A kitchen harvest of living lettuce. Amer. Veg. Grower 29(11); 10,12.

Sheldrake, R. and R. Langhans. 1961. Heating study with plastic greenhouses. Proc. Nat. Hort. Plastics Cong. 2:16-17.

Shimokawa, A., and H. Ono, 1954. Studies on the forcing of cucumber culture in vinyl film tunnels. Bul. Hort. Branch Kanagawa Agric. Exp. Sta. 2:33-40.

Skaggs, R. W., D. C. Sanders, and C. R. Willey. 1976. Use of waste heat for soil warming in North Carolina. Trans. Amer. Soc. Agr. Eng. 19:159-167.

Sonneveld, C. 1988. Rockwool as a substrate in protectd cultivation. Hort. In High Techn. Eva. Tokyo, Japan. p.173-191.

Strickler, P. M. 1975. The use of plastic for heat insulation in greenhouses. Plasticulture 25: 41-53.

Sturbey, N. 1980. Station aims for more flexible NFT cropping. Grower 94(4):8.

Takakura, Tadashi. 1988. Protected cultivation in Japan. Symp. on High Tech. in Protected Cult. Acta Hort. 230:29-37.

Tarakanov, G. I. and N. F. Rozov. 1962. The improvement of micro-climate under plastic tunnels on unheated ground. Dobl. Mosk. sel'doz. Akad. K. A. Timirjazera, No. 77. pp. 297-306.

Tayama, H. K. and T. J. Roll, Ed. 1989. Tips on growing bedding plants. Ohio Cooperative Extension Service, Ohio State Uni., Columbus, Ohio.

Tayama, H. K. and T. J. Roll. Ed. 1989. Tips on gowing potted chrysanthemums. Ohio Cooperative Ext. Service, Ohio State Uni., Columbus, Ohio

Thompson, H. C. and W. C. Kelly. 1957. Vegetable Crops. Fifth Edition. McGraw-Hill Book Co., Inc. NY. Chapt. 7:86-106.

Toms, B. and B. MacPhail. 1991. Greenhouse tomato/cucumber trials. Nova Scotia Greenhouse Newsletter No. 3, Plant Ind. Branch, Nova Scotia Dept. of Agric., Truro.

Tzekleev, G., L. Guzelev, Y. Solakov and S. Stoilov. 1988. Plastics in The People's Republic of Bularia. Plasticulture 78:19-27.

Van Os, E.A. 1983. Dutch developments in soil s culture. Outlook in Agr.

Varley, M.J. and S. W. Burrage. 1981. New solution for lettuce. Grower 95(l5):19-21, 23, 25.

Vogal, G. 1963. The effect of short-term covering with plastic tunnels on the croppings of some vegetables grown in spring and summer. Arch. Gartenb. 11:27-46.

Wardlow, L. R., F. A. B. Ludlam, and N. French. 1972. Insecticide resistance in greenhouse whitefly. Nature (London) 239:164-165.

Wells, O.S. and J. B. Loy. 1980. Slitted row covers for intensive vegetable production. Coop. Ext. Service, Uni. of N.H. p. 4.

Wells, O.S. and J. B. Loy. 1985. Intensive vegetable production with row covers. HortScience 20(5): 822-826.

Wells, O. S., J. B. Loy, and T. A. Natti. 1977. Slit mulch film used as row covers. Proc. Nat'l. Agric. Plast. Cong. 13:448-452.

Wells, O.S. 1991. High tunnels shelter early crops. Amer. Veg. Grow. 39(2):44-47.

Wheatley, J. C. 1991. Row covers act as insect barrier. Amer. Veg. Grower, 39(4): 21-22.

White, H. 1980. NFT in the air makes for easier management. Grower 94(2):28,31-32.

White, J. W., R. A. Aldrich, K. Vedam, J. L. Duda, S. M. Rebuck, G. R. Mariner, and J. R. Smith. 1976. Energy conservation systems for greenhouses. Proc. Solar Energy Fuel-Food Workshop. Uni. of Arizona, Tucson. p. 191-212.

Winsor, G. W., R. G. Hurd, and D. Price. 1979. Nutrient Film Technique. Grower Bulletin. 5. Glasshouse Crops Research Institute, Littlehampton, England.

Wittwer, S.H. 1981. The 20 crops that stand between man and starvation. Farm Chemicals. 8:17-18, 23, 26, 28.

Woodbury, G. W. l963. Research shows muskmelons grow better under plastic mulch. Idaho Agric. Sci. 48(3):3.

GLOSSARY

Compiled by: James E. Brown and Lisa B. Jacks
Department of Horticulture, Auburn University, Alabama

Reprinted by permission of authors.

ABRASION - Damage caused by scuffing or friction.

ABSORPTANCE - Ability of a material to absorb energy.

ACRYLIC (or acrylates) - Thermoplastic formed from polymers of acrylic acid or its derivatives, particularly methyl methacrylate, known for it's light transmission and weather resistance.

ACRYLONITRILE - A monomer with a structure (CH_2:CHCN). It is most useful in copolymers. It's copolymer with butadiene is nitrile rubber. It is also used to make synthetic fiber and as a chemical intermediate.

ACRYLONITRILE-BUTADIENE-STYRENE (Abbreviated ABS) - Acrylonitrile and styrene liquids and butadiene gas are polymerized together in a variety of ratios to produce the family of ABS resins.

ADIABATIC - An adjective used to describe a process or transformation in which no heat is added or allowed to escape from the system under consideration. It is used, somewhat incorrectly, to describe a mode of extrusion in which no external heat is added to the extruder although heat may be removed by cooling to keep the output temperature of the melt passing through the extruder constant. The heat input in such a process is developed by the screw as its mechanical energy is being converted into thermal energy.

AGEING - Changes in a material due to time and environmental conditions which degrade or improve it.

AIR MASS - Path length of radiation through the atmosphere, considering the vertical path at sea level as unity.

ALLOY - Composite material made up by blending polymers or copolymers with other polymers or elastomers under selected conditions, e.g., styrene-acrylonitrile copolymer resins blended with butadieneacrylonitrile rubbers.

ANTIOXIDANT - Formulation ingredient which prevents or slows down oxidation of plastic material exposed to air.

ANTISTATIC AGENTS - Methods of minimizing static electricity in plastic materials. Such agents are of two basic types: (1) metallic devices which come into contact with the plastics and conduct the static to earth. Such devices give complete neutralization at the time, but because they do not modify the surface of the material it can become prone to further static during subsequent handling; (2) emical additives which, mixed with the compound during processing, give a reasonable degree of protection of the finished products.

ARTIFICIAL AGEING - The accelerated testing of plastics specimens to determine their changes in properties. Carried out over a short period of time, such tests are indicative of what may be expected of a material under service conditions over extended periods. Typical investigations include those for dimensional stability; the effects of immersion in water, chemi-

cals and solvents; light stability and resistance to fatigue.

ASTM - American Society for Testing Materials.

AVERAGE MOLECULAR WEIGHT (Viscosity method) - The molecular weight of polymeric materials determined by the viscosity of the polymer in solution at a specific temperature. This gives an average molecular weight of the molecular chains in the polymer independent of specific chain length. The value falls between weight average and number average molecular weight.

BEAM RADIATION - Solar radiation received from the sun without change of direction.

BLACKBODY - By definition, a perfect absorber and emitter of all radiation. The perfect blackbody will absorb and emit all incident radiation regardless of wave length or direction.

BLEED - To give up color when in contact with water or a solvent; undersized movement of certain materials in a plastic (e.g. plasticiers in vinyl) to the surface of the finished article or into an adjacent material. Also called Migration.

BLISTER - A raised arc on the surface of a molding caused by the pressure of gases inside on its incompletely hardened surface. IN FRP this is a dull surface area created by the breaking of the bond between the cellophane/polyester/polyvinyl fluoride film and the laminate. This is the direct result of excess vapors trying to escape during the monomer ross-linking.

BLOCKING - An undesired adhesion between touching layers of a material, such as occurs under moderate pressure during storage production or use.

BLOOMING - The process of material components migrating to the surface.

BLOW MOLDING - The process of forming hollow articles by expanding a hot plastic in the internal surfaces of a mold.

BLOWN TUBING (Blown film) - A thermoplastic film which is produced by extruding a tube, applying a slight internal pressure to the tube to expand it while still molten and subsequent cooling to set the tube. The tube is then flattened through guides and wound up flat on rolls. The size of blown tubing is determined by the flat width in inches as wound rate than by the diameter as in the case of the various rigid types of tubing.

BREATHING - When referring to plastic sheeting, "breathing" indicates permeability to air.

BURNING RATE - A term describing the tendency of plastics articles to burn at given temperatures. Certain plastics, such as those based on shellac, burn readily at comparatively low temperatures. Others will melt or disintegrate without actually burning, or will burn only if exposed to direct flame. These latter are often referred to as self-extinguishing, the relative speed that plastics burn under given conditions.

CALENDAR (v.) - To prepare sheets of materials by pressure between two or more counter-rotating rolls. (n.) - The machine performing this operation.

CALIPER - Thickness or gauge, usually expressed in mils (thousandths of an inch).

CARBON BLACK - A black pigment produced by the incomplete burning of natural gas or oil. It is widely used as a filler, particularly in the rubber industry. Because it possesses useful ultraviolet protective properties, it is also much used in polyethylene compounds intended for such applications as cold water piping and black agricultural sheet.

CELLULOSE - A natural high polymeric carbohydrate found in most plants; the main constituent of dried woods, jute,flax, hemp, ramie, etc. Cotton is almost pure cellulose.

CLARITY - Freedom from haze; transparency.

COEFFICIENT OF EXPANSION - The fractional change in length (sometimes volume, specified) of a material for a unit change in temperature. Values for plastics range from 0.01 to 0.2 mils/in. $^{\circ}$C. This is the number of mils expanded for each inch of total length when the temperature rises one degree centigrade.

COMPOSITE - A plastic structure consisting of two or more different materials intimately mixed together. (FRP is an example)

CONDENSATION STRENGTH - Crushing load at the failure of a specimen divided by the original sectional arc of the specimen.

CONDENSATION - A chemical reaction in which two or more molecules combine with the separation of water or some other simple substance. If a polymer is formed, the condensation process is called Polycondensation. See also Polymerization.

CONDITIONING - The subjection of a material to a stipulated treatment so that is will respond in a uniform way to subsequent testing or processing. The term is frequently used to refer to the treatment given to specimens before testing.

COPOLYMER - A polymer of two or more chemically distinct monomers.

CRAZING - Fine cracks which may extend in a network on or under the surface or through a layer of a plastic material.

CREEP - A change in shape of material under load over a period of time. Creep at room temperature is sometimes called Cold Flow.

CROSS-DIRECTION or "C.D." - "Transverse Direction."

CROSS-LINKING - Applied to polymer molecules, the setting-up of chemical links between the molecular chains. When extensive, as in most thermosetting resins, cross-linking makes one infusible super-molecule of all the chains.

DEFLECTION - The bending or sagging of a material because of a load.

DEGREE DAY - A unit based on temperature differences and time used in estimating fuel consumption and specifying nominal annual heating load of a building.

DELAMINATION - The separation of a material into layers or sheets.

DENIER - The weight (in grams) of 9000 meters of synthetic fiber in the form of continuous filament.

DENSITY - The weight of a material per cubic unit. Pounds per cubic foot, grams per cubic centimeter, etc.

DESTATICIZATION - Treating plastics materials to minimize their accumulation of static electricity and, consequently, the amount of dust picked up by the plastics because of such charges.

DIE LINES - Vertical marks on the parison caused by damage of die parts or contamination.

DIE-CUT - Punched out or cut in a pattern by means of a sharp tool.

DIFFUSE RADIATION - Solar radiation received from the sun after the radiation's direction has been changed by reflection and scattering.

DIMENSIONAL STABILITY - The ability of a plastic part to retain the precise shape in which it was originally formed under changing conditions of temperature and humidity.

DISCOLORATION - Any change from the original color, often caused by overheating, light exposure, irradiation, or chemical attack.

DISPERSION - The finely divided particles of a material in suspension in another substance.

DUROMETER HARDNESS - The hardness of a material as measured by the Shore Durometer.

ELASTIC DEFORMATION - The temporary change in shape caused by a load which is recoverable when the load is removed.

ELASTICITY - That property of a material by virtue of

which it tends to recover its original size and shape after deformation. The tendency of a material to recover its natural size and shape when deforming load is removed.

ELASTOMER - A material which at room temperature stretches under low stress to at least twice its length and then snaps back to its original length upon release of the applied stress.

ELECTRONIC TREATING - A method of oxidizing a film of polyethylene to render it printable by passing the film between the electrodes and subjecting it to a high voltage corona discharge.

ELMENDORF TEST - A method of testing film resistance to tear, by which the weight required to tear one or several notched layers of film under test is measure. Results are usually reported in grams for both machine's direction and across the web.

ELONGATION - The increase in length of a material being loaded in tension.

EMBOSSING - Techniques used to create depression of a specific pattern in plastic film and sheeting.

EMITTANCE - The giving off of energy by a material.

EMULSION - A suspension of fine droplets of one liquid in another.

ENVIRONMENTAL STRESS CRACKING (ESC) - The appearance of network of fine cracks on or under the surface of a thermoplastic caused by the exposure to weather.

ETHYLENE-VINYL ACETATE - Copolymers from these two monomers form a new class of plastic materials. They retain many of the properties of polyethylene, but have considerable increased flexibility for their density - elongation and impact resistance are also increased.

EXTRUSION - Forcing molten plastic material through a die to form a continuous shaped article.

FABRICATE - To work a material into a finished form by machining, forming, or other operation or to make flexible film or sheeting into end products by sewing, cutting, sealing, or other operation.

FIBERGLASS REINFORCED PLASTIC (FRP) - Low pressure thermosetting or thermoplastic laminate consisting of a resin system strengthened by reinforcing strands of fiberglass.

FILAMENT WINDING - Roving or single strands of glass, metal, or other reinforcement are wound in a predetermined pattern onto a suitable mandrel. The pattern is so designed as to give maximum strength in the directions

required. The strands can either be run from creel through resin bath before winding or preimpregnated materials can be used. When the right number of layers has been applied, the wound mandrel is cured at room temperatures or in an oven.

FILLER - A cheap, inert substance added to a plastic to make it less costly. Fillers may also improve physical properties, particularly hardness, stiffness, and impact strength. The particles are usually small, in contrast to those of reinforcement (q.v.), but there is some overlap between the functions of the two.

FILM - A thin skin, layer, sheet or coating with a thickness less than 0.010 inch.

FILM, NON-FOGGING - Film which does not become cloudy from condensation of moisture caused by temperature drops or humidity changes.

FILM, ORIENTED - Film in which the molecular structure is aligned mechanically in one or more directions, giving the film more strength while introducing shrinkage characteristics.

FISH EYE - A fault in transparent or translucent plastics materials, such as film or sheet, appearing as a small globular mass and caused by incomplete blending of the mass with surrounding material.

FLAME-RESISTIVE - Commonly, that property of a material or combination of materials that will improve its' ability to resist ignition when exposed to flame.

FLAME RETARDANT RESIN - A resin which is compounded with certain chemicals to reduce or eliminate its tendency to burn. For polyethylene and similar resins, chemicals such as antimony trioxide and chlorinated paraffins are useful.

FLAME SPREAD CLASSIFICATION - A relative number, based on comparison with red oak and cement-asbestos, of the time elapsed and distance traveled by a flame front on the surface of a specimen tested in accordance with ASTM E-84 (Tunnel Test—also designated UL 723, UBC Standard 42-1 etc.).

FLAME TREATING - A method of rendering inert thermoplastic objects receptive to inks, lacquers, paints, adhesives, etc., in which the object is bathed in an open flame to promote oxidation of the surface of the article.

FLAMMABILITY - A measure of a material's ability to support combustion, usually determined by one or more small-scale laboratory tests.

FLASH POINT - The lowest temperature at which a combustible liquid will give off a flammable vapor that will burn momentarily.

FLEXURAL STRENGTH - The strength of a material

in bending, expressed as the tensile stress of the outer most fibers of a bent test sample at the instant of failure. With plastics, this value is usually higher than the straight tensile strength.

FOOT CANDLE - The amount of illumination obtained when a source of 1 candle power illuminates a screen 1 foot away.

FORMING - A process in which the shape of plastic pieces such as sheets, rods, or tubes is changed to a desired configuration. The use of the term forming in plastics technology does not include such operations as molding, casting, or extrusion, in which shapes or articles are made from molding materials or liquids.

FRICTION WELDING - A method of welding thermoplastics materials whereby the heat necessary to soften the components is provided by friction.

GAS TRANSMISSION - The movement of gas through film materials. The gas transmission property (permeability) of a film is measured in terms of volume of gas (at standard temperature and pressure) transmitted through a given area of film of a given thickness within a given time.

GAS TRANSMISSION RATE - A measure of the permeability of a film to gases.

GEL - In polyethylene, a small amorphous resin particle which differs from its surroundings by being of higher molecular weight and/or cross-linked, so that its processing characteristics differ from the surrounding resin to such a degree that it is not easily dispersed in the surrounding resin. A gel is readily discernible in thin films.

GEL COAT - A thin, surface layer of resin, sometimes with pigment, applied to a reinforced plastics panel or molding. It provides additional surface weather resistance, and improved appearance.

GLOSS - Term used to express shine, sheen, or lustre of a film surface.

GUSSET - A tuck placed in each side of a tube of blown tubing as produced to provide a convenient square or rectangular package, similar to that of the familiar brown paper bag or sack, in subsequent packaging.

HARDNESS - The resistance of a plastics material to compression and indentation. Among the most important methods of testing this property are Brinell hardness, Rockwell and Shore hardness.

HAZE - The cloudiness in a transparent material or the portion of light that is not transmitted through a material in a straight line.

HEAT DISTORTION TEMPERATURE - The temperature at which a material starts to lose its mechanical properties and easily becomes distorted.

HEAT-SEALING - A method of joining plastic films by simultaneous application of heat and pressure to areas in contact. Heat may be supplied conductively or dielectrically.

HEAT TRANSMISSION - The allowing of heat (energy) to pass through a substance.

HERMETIC SEAL - A seal that will exclude air and be leak-proof.

HOMOGENEOUS - Uniform structure or composition through the material.

HOMOPOLYMER - A polymer consisting of (neglecting the ends, branch junctions and other minor irregularities) a single type of repeating unit.

HORIZONTAL/VERTICAL BURN TESTS - For materials which are entirely consumed in such tests, a rate of burning (centimeters or inches per second) value is reported. For those materials which do not continue to burn, two values are reported; i.e. Average Extent of Burning (AEB-burn length) and an Average Time of Burning (ATB). These small scale tests measure combustibility. Lower values indicate greater resistance to burning.

HYGROSCOPIC - Having the property of readily absorbing moisture from the atmosphere.

I.D. - Inside diameter.

IGNITION TEMPERATURE - The temperature at which a material will start burning without an external flame (Self Ignition Temperature). Higher values indicate greater resistance to burning.

IMPACT STRENGTH - The ability to resist breakage by flexural shock.

INDEX OF REFRACTION - The ratio of the velocity of light in the first of two media to its velocity in the second as it passed from one to the other. The closer the ratio, the more visual similarity of the media.

INHIBITOR - A substance that slows down chemical reaction. Inhibitors are sometimes used in certain types of monomers and resins to prolong storage life.

INJECTION BLOW MOLDING - A blow molding process in which the parison to be blown is formed by injection molding.

INJECTION MOLDING - A molding procedure whereby a heat-softened plastic material is forced from a cylinder into a relatively cool cavity which gives the article the desired shape.

INSOLATION - acronym for INCOMING SOLAR RADIATION - a term describing the level or intensity of sunlight, measured in Langleys.

IZOD IMPACT TEST - A test that measures the ability of a material to withstand impact (instant loading). The specimen is held in jaws and struck with weighted pendulum.

"K" FACTOR-(THERMAL CONDUCTIVITY) - This factor indicates the amount of heat in BTU's transferred in one hour through one square foot of homogeneous material l" thick, with a temperature difference of one degree F between its two surfaces.

LAMINATED PLASTIC - Multiple layers of plastic bonded together.

LANGLEY - A unit of solar radiation equivalent to one gram calorie per square centimeter of irradiated surface.

LIGHT-RESISTANCE - The ability of a plastic material to resist fading after exposure to sunlight or ultraviolet light. Nearly all plastics tend to darken under these conditions.

LIGHT TRANSMISSION - The percent of total available light that passes through a material.

MACHINE DIRECTION - In the manufacture of film the majority of the molecules tend to align themselves in a direction parallel to the direction of travel through the machine.

MANOMETER - An instrument for measuring pressure.

MEGA-JOULE - A unit of measurement of energy.

MELT INDEX - The amount, in grams, of a thermoplastic resin which can be forced through a 0.0825 inch orifice when subjected to 3160 grams force in 10 minutes at 190 degrees C.

MELT STRENGTH - The strength of the plastic while in the molten state.

METALLIZING - Applying a thin coating of metal to a nonmetallic surface. May be done by chemical deposition or by exposing the surface to vaporized metal in a vacuum chamber.

MICROMETER CALIPER - An instrument for measurements in terms of minute dimensions, usually in 0.001" or 0.0001".

MICRON - A unit of length equal to 1/1000 of a millimeter.

MIL - A unit of length equal to 0.001 inch, often used for reporting film thickness. To convert mil to millimeter, multiply by 0.0254.

MODIFIED - Containing ingredients such as fillers, pigments or other additives, that help to vary the physical properties of a plastics material. An example is oil modified resin.

MODULUS OF ELASTICITY - The ratio of stress to strain in a material that is elastically deformed.

MOISTURE VAPOR TRANSMISSION - The rate at which water vapor permeates through a plastic film or wall at a specified temperature and humidity.

MONOMER - A relatively simple compound which can react to form a polymer. See also polymer.

NANOMETER - A unit of length equal to 1/1000 of a micron.

NONWOVEN - A textile made of fibers held together by interlocking or bonding (e.g. spunbonded, meltblown, bonded carded web).

O.D. - Outside diameter.

OPACITY - Resistance of material to transmission of light.

OPAQUE - Descriptive of a material or substance which will not transmit light. Opposite of transparent, q.v. Materials which are neither opaque nor transparent are sometimes described as semi-opaque, but are more properly classified as translucent, q.v.

ORIENTATION - The alignment of the crystalline structure in polymeric materials so as to produce a highly uniform structure by cold drawing or stretching during fabrication.

OXIDATION - Reaction of a substance with oxygen. The breakdown of the surface of a material due to prolonged exposure to oxygen.

PARISON - The hollow plastic tube from which an item is blow molded.

PERMEABILITY - 1) The passage or diffusion of a gas, vapor, liquid, or solid through a barrier without physically or chemically affecting it. 2) the rate of such passage.

PHOTOSYNTHETICALLY ACTIVE RADIATION - The P.A.R. range is that portion of sunlight ranging in wavelength from 400 to 700 nanometers. This waveband is also defined as visible light, i.e. those wavelengths seen by the human eye.

PINHOLE - A very small hole in an extruded resin coating of film.

PITCH - The distance between centers of two adjacent ribs or corrugations.

PLASTIC - Any of numerous organic, synthetic, or processed materials that are molded, cast extruded, drawn, or laminated into objects, films, or filaments. At some stage of the manufacturing process, all plastics are capable of flowing, under heat and pressure, if necessary, into the desired final shape.

PLASTIC DEFORMATION - A change in dimension of an object under load that is not recovered when the load in removes; opposed to elastic deformation.

PLASTICIZE - To soften a material and make it plastic or moldable, either by means of a plasticizer or the application of heat.

PLASTICIZER - Chemical agent added to plastics to make them softer and more flexible.

PLASTICIZER MIGRATION - Undefined movement of the plasticizer to the surface of a plastic, or from one plastic to another, or from a plasticized substance into the atmosphere.

POLYESTER - A longchain high-molecular-weight resin commonly used as base for glass-fiber-reinforced plastic.

POLYCARBONATE - One of the many thermoplastics, polycarbonate is known for its clarity and toughness.

POLYETHYLENE - There are two major types with different chain structures: The stiffer, stronger, linear material, sometimes called high-density (HDPE) or low-pressure and the more flexible, lower-melting, branched polyethylene, known as low-density (LDPE) or high pressure polyethylene. More recently,linear low-density polyethylene (LLDPE) became a major product.

POLYMER - A high-molecular-weight organic compound, natural or synthetic, whose structure can be represented by a repeated small unit, the mer; e.g., polyethylene, rubber, cellulose. Synthetic polymers are formed by addition or condensation polymerization of monomers. If two or more monomers are involved, a copolymer is obtained. Some polymers are elastomers, some plastics. A long-chain compound made up of many repeating units joined together end to end.

POLYMERIZATION - A chemical reaction in which the new molecule's molecular weight is a multiple of that of the original substance. When two or more monomers are involved, the process is called copolymerization or heteropolymerization.

POLYPROPLYENE - One of many thermoplastics, polypropylene is known for its versatility and toughness.

POLYSTRENE - A thermoplastic material derived from the polymerization of styrene monomers.

POLYURETHANE - Used in foams, fibers, elastomers, and coatings.

POLYVINYLCHLORIDE (PVC) - A plastic polymer of the monomer vinyl chloride. It is a rigid clear thermoplastic and has good impact resistance. A colorless solid with outstanding resistance to water, alcohols, and concentrated acids and alkalies. It is obtainable in the form of granules, solutions, latices, and pastes. Compounded with plasticizers, it yields a flexible material superior to rubber in aging properties. It is widely used for cable and wire coverings, in chemical plants, and in the manufacture of protective garments.

POLYVINYLFLUORIDE - A film-forming thermoplastic with good weather properties. (Tedlar).

PREPREG - A term generally used in reinforced plastics to mean the reinforcing material containing or combined with the full complement of resin before molding.

PSI - Pounds per square inch.

PURLINS - A horizontal member in a roof supporting the common rafters.

PYRANOMETER - Instrument for measuring the total hemispherical solar radiation, usually on a horizontal surface. Measures both beam and diffuse radiation.

QUON-SET - Trademark for prefabricated shelter set on foundation of bolted steel trusses and build of a semicircular arching roof.

RECYCLE - The ground material from flash and trimmings which, after mixing with a certain amount of virgin material, is fed back into the molding machine.

REFRACTION - Passage of light from air to a denser optical medium which causes a reduction in velocity and direction. The ratio of these velocities is known as the refractive index of the denser medium.

REFLECTANCE - Energy that has not been transmitted or absorbed.

REINFORCEMENT - A strong material added to a base material to improve its strength, stiffness, and impact resistance. To be effective, the reinforcing material must form a strong bond with the base material.

RESILIENCY - Ability to quickly regain an original shape after being strained or distorted.

RESIN - A solid or semi-solid, complex mixture of organic compounds. Usually a polymer. Sometimes used interchangeably with the word "plastic."

"R" FACTOR (RESISTANCE) - The reciprocal of conductivity (K) or conductance (C). Divide the known K or C factor into 1.0 -the answer is the R factor.

RIB - A stiffening member of a fabricated part. Having a straight section as opposed to arched corrugations.

RIGID PVC - Polyvinyl chloride or a polyvinyl chloride/acetate copolymer characterized by a relatively high degree of hardness; it may be formulated with or without a small percentage of plasticizer.

ROCKWELL HARDNESS - A common method of testing a plastics material for resistance to indentation in which a diamond or steel ball, under pressure, is used to pierce the test specimen. The load used is expressed in kilograms and a 10-kilogram weight is first applied and the degree of penetration noted. The so-called major load (60-150 kilograms) is next applied and a second reading obtained. The hardness is then calculated as the difference between the two loads and expressed with nine different prefix letters to denote the type of penetrator used and the weight applied as the major load.

ROCKWOOL - Mineral wool made by blowing a jet of steam through molten rock (as limestone or siliceons rock) or through slag.

ROVING - A form of fibrous glass in which spun strands are woven into a tubular rope. The number of strands is variable but 60 is usual.

SELF-EXTINGUISHING - A somewhat loosely-used term describing the ability of a material to cease burning once the source of flame has been removed.

SHADING COEFFICIENT - The ratio of solar heat gain through a glazing system under a specific set of conditions, to the solar heat gain through a single pane of double-strength sheet glass under the same set of conditions.

SHEAR STRENGTH - The ability of a material to resist shear stress. The maximum load required to shear the specimen in such a manner that the moving portion has completely cleared the stationary portion.

SHEET (THERMOPLASTIC) - A flat section of a thermoplastic film with the length considerably greater than the width and 10 mils (0.01) or greater in thickness.

SHRINK WRAPPING - A technique of packaging in which the strains in a plastic film are released by raising the temperature of the film thus causing it to shrink over the package. These shrink characteristics are built into the film during its manufacture by stretching it under controlled temperatures to produce orientation, q.v., of the molecules. Upon cooling, the film retains its stretch condition, but reverts toward its original dimensions when it is heated. Shrink film gives good protection to the products packaged and has excellent clarity.

SLIP ADDITIVE - A modifier that acts as an internal lubricant which exudes to the surface of the plastic during and immediately after processing. In other words, a nonvisible coating blooms to the surface to provide the necessary lubricity to reduce coefficient of friction and thereby improve the slip characteristics.

SLOT EXTRUSION - A method of extruding film sheet in which the molten thermoplastic compound is forced through a straight slot.

SMOKE DENSITY - The amount of smoke generated while a material is burning.

SOFTENING RANGE - The range of temperature in which a plastic changes from a rigid to a soft state. Actual values will depend on the test method. Sometimes erroneously referred to as softening point.

SOLAR ENERGY TRANSMISSION - The percentage of total solar energy which will pass through a material.

SOLAR INFRARED ENERGY - That portion of sunlight ranging from 700 to 2500 nanometers wavelengths, often called "near" infrared. "Far" infrared, or "black body" radiation, ranges from 2500 to 10,000 nm.

SOLAR RADIATION - The entire energy spectrum (all wave lengths) created by the sun. The wavelengths reaching earth range from 300 nm to 3000 nm.

SOLVENT - A liquid or agent that dissolves another substance.

SPAN - The distance between supports such as rafter or purlins that a material must cross.

SPANISHING - A method for depositing ink in the valleys or embossed plastics film.

SPECIFIC GRAVITY - The density (mass per unit volume) of a material divided by the density of water. Since the density of water is 1.0 gram/cubic centimeter, density in grams per cubic centimeter is numerically equal to specific gravity.

SPECTRUM - An array of components of an emission separated and arranged in the order of some varying characteristic. (Wave lengths).

SPLICE - To unite or join the end of roll material. The bond may be accomplished by mechanical or electrical means, or by an adhesive.

SPUNBONDED - A nonwoven textile formed of extruded continuous high denier fibers and bonded by thermal, chemical, or mechanical means.

STABILIZER - An ingredient used in the formulation of some plastics, especially elastomers, to assist in maintaining the physical and chemical properties of compounded material at their initial values throughout processing and service lives of the material.

STORAGE LIFE - The period of time during which a liquid resin or packaged adhesive can be stored under specified temperature conditions and remain suitable for use. Storage life is sometimes called Shelf Life.

STRESS CRACK - External or internal cracks in a plastic caused by tensile stresses less than that of its short-time mechanical strength. The development of such cracks is frequently accelerated by the environment to which the plastic is exposed. The stresses which cause cracking may be present internally or externally or may be combinations of the stresses. The appearance of a network of fine cracks is crazing.

SURFACE TREATING - Any method of treating a polyolefin so as to alter the surface and render it receptive to inks, paints, lacquers and adhesives such as chemical, flame, and electronic treating.

TENSILE STRENGTH - The pulling stress, usually expressed in psi, required to break a given specimen.

THERMAL CONDUCTIVITY - The ability of a substance to let heat pass through a unit cube of the material. (See K-Factor).

THERMAL RADIATION - That portion of the spectrum encompassing all of the infrared and visible portions of the spectrum, as well as a small portion of the ultra-violet, approximately 0.3 to 1,000 microns.

THERMOFORMING - Any process of forming thermoplastic sheet which consists of heating the sheet and pulling it down onto a mold surface.

THERMOPLASTIC - (a) Capable of being repeatedly softened by heat and hardened by cooling. (b) A material that will repeatedly soften when heated and harden when cooled. Typical of the thermoplastics family are the styrene polymers and copolymers, acrylics, cellulosics, polyethylenes, vinyls, nylons, and the various fluorocarbon materials.

THERMOSET - A material that will undergo or has undergone a chemical reaction by the action of heat, catalysts, ultraviolet light, etc., leading to a relatively infusible state. Typical of the plastics in the thermosetting family are the aminos (melamine and urea), most polyesters, alkyds, epoxies, and phenolics.

TOLERANCE - A specified allowance for deviations in weighing, measuring, etc., or for deviations from the standard dimensions or weight.

TOXIC - Poisonous.

TRANSLUCENT - Descriptive of material or substance capable of transmitting some light, but not clear enough to be seen through.

TRANSMISSION - The amount of energy a material allows to pass through itself.

TRANSPARENT - Descriptive of a material or substance capable of a high degree of light transmission (e.g., glass). Some polypropylene films and acrylic moldings are outstanding in this respect.

"U" FACTOR - The total or overall transmission of heat in one hour per square foot of area that will pass through a combination of materials and air spaces. (Expressed in BTU's).

ULTIMATE STRENGTH - Term used to describe the maximum unit stress a material will withstand when subjected to an applied load in a compression, tension, or shear test.

ULTRASONIC SEALING - A film sealing method in which sealing is accomplished through the application of vibratory mechanical pressure at ultrasonic frequencies (20 to 40 kc.). Electrical energy is converted to ultrasonic vibrations through the use of either a magnetostrictive or piezoelectric transducer. The vibratory pressures at the film interface in the sealing area develop localized heat losses which melt the plastic surfaces, effecting the seal.

ULTRAVIOLET - Zone of invisible radiations beyond the violet end of the spectrum of visible radiations. Since UV wavelengths are shorter than the visible, their photons have more energy, enough to initiate some chemical reactions and to degrade most plastics.

ULTRAVIOLET LIGHT - That portion of sunlight ranging in wavelength from 200 to 400 nanometers. Wavelengths below 300 nm do not pass through the atmosphere in appreciable amounts.

"U" VALUE - Overall rate of heat transfer through a material. "R-value" (1/U value), resistance to heat transfer is more commonly used in consumer advertising. These values are calculated by the ASHRAE method.

UV ABSORBERS - A chemical compound with the ability to selectively absorb UV radiation. When incorporated into plastics it reduces the degrading effects of ultraviolet.

UV STABILIZER (ULTRAVIOLET) - Any chemical compound which, admixed with a thermoplastic resin, selectively absorbs UV rays.

VERTICAL/HORIZONTAL BURN TEST - See Horizontal.

VISIBLE LIGHT - The band of colors making up white light. All radiation that can be sensed by the human eye. Approximately 0.38 to 0.78 microns.

WATER VAPOR PERMEABILITY - The ability of a material to permit transmission of water vapor.

WATER VAPOR TRANSMISSION RATE (WVTR) - Measure of permeability of a material, often stat ed in terms of grams of water passing through 100 square inches of material in 24 hours at 100 degrees F and 90% humidity.

WEATHERING - The effects of outdoor exposure on a material, including water, UV radiation, oxidation, etc.

WEATHEROMETER - An instrument used to subject materials to accelerated weathering conditions. Example: Rich UV source and water spray.

WINDOW - A defect in a thermoplastics film, sheet, or molding, caused by the incomplete "Plasticization" of a piece of material during processing. It appears as a globule in an otherwise blended mass. See also Fish Eye.

YIELD - Area per unit of weight, usually expressed as square inches per pound.

YOUNG'S MODULUS OF ELASTICITY - Stress divided by strain.

ANNEX

MARKETING AND DISTRIBUTION OF PROTECTED CROPS

TABLE OF CONTENTS - ANNEX

1

INTRODUCTION

As described in the main text of "A Global Review of Protected Agriculture," almost any horticultural crop can be grown under some form of protective cover, thereby increasing production, improving quality and extending seasonal availability. Such produce primarily consists of seasonally produced, non-storable products. The large and expanding northern hemisphere markets appear to offer plentiful export opportunities for many developing countries with complementary growing seasons. Countries such as Morocco, which supplies fresh vegetables, and Colombia and Kenya, which produce fresh cut flowers, illustrate how suppliers with complementary growing seasons may successfully establish themselves in the northern hemisphere market.

Despite such specific successes, imports from external suppliers into northern hemisphere countries remain relatively limited. The more developed countries still supply the overwhelming proportion of their own produce needs, even during their "off-seasons": in 1990, imports accounted for only 16% of the market value of cut flowers in the United States (U.S.), 7% in the European Community (EC) and about 5% in Japan (Rabobank-F). In mid-winter, about 75% of the cut flowers sold in the Dutch auctions—the major intermediate market for floral supplies to the whole of Europe—are grown in the Netherlands (VBN). Similarly, 86% of the fresh vegetable imports into EC countries in 1990 came from other EC countries and only 5% came from the Canary Islands (now part of the EC), 4% from Morocco and 5% from other non-EC countries (Rabobank-V).

This section discusses the marketing of protected agricultural crops, primarily fresh vegetables and fresh cut flowers. The first half of the section examines the complications of horticultural marketing which arise from the high degree of perishability of these products, the infinite combination of possible production variables, an increasingly blurred concept of seasonality, the critical importance of quality and consumer preferences, and the need for comprehensive and timely information about production and market developments. The second half of the section examines the basic trends, opportunities and trade policy environment in the horticultural markets of the United States, Western Europe, and Japan.[2]

WHAT ARE "PROTECTED CROPS"?

From a marketing standpoint, all vegetables and cut flowers for fresh consumption will be considered "protected crops", since many of these products are grown with some degree of protective covering, and since the production volume and value of such crops is much greater than the total covered production area would indicate. In developed countries, vegetable production under glass or plastic cover is estimated to be about 7% of the total vegetable production area, ranging from as much as 10% in the Netherlands (4,420 ha.), to 5% in Belgium (2,535 ha.), 2% in Italy (9,000 ha.), 1% in Poland (3,300 ha.), and much less than 1% in the U.S. (only about 320 ha.). However,

since covered production results in much higher productivity and quality, glasshouses in the Netherlands account for 52% of the total quantity and 75% of the total value of Dutch vegetables—on only 10% of the total production area. In many countries, covered production also predominates for most salad vegetables which are grown for fresh consumption, such as tomatoes, cucumbers, peppers and lettuce (Rabobank-V, Neff).

A much larger proportion of the world's cut flower growing area (approximately 30%) is under cover than is true for vegetable production. The figure is considerably higher in northern countries and in countries producing primarily for export, and lower in countries with mild year-round climates and in those producing mainly for the domestic market. Coverage of the production area ranges from almost 100% in Colombia (4,000 ha.) to 71% in the Netherlands (4,050 ha.), 65% in Israel (1,200 ha.), 61% in Spain (1,400 ha.), 54% in Italy (4,130 ha.), 40% in the U.S. (8,000 ha., including pot plants), 33% in Japan (5,500 ha.), and 10% in Kenya (55 ha.).

Some countries, such as Germany, are expanding their flower growing area in the open, because growers believe they cannot afford the high cost of glasshouses. By contrast, the area under cover is increasing in countries such as Italy, Spain and Colombia because growers in these countries feel that they cannot produce high quality products and thus compete in northern markets unless they cover their crops. Still other countries, such as the Netherlands, Israel and Kenya, are increasing both covered and open growing areas as part of a general expansion of their floriculture sectors (Rabobank-F, USDA-NASS). In Israel, for example, crops which require greater light intensity, such as waxflowers and gladiolas, can be grown more extensively and without cover (but with irrigation); most other flowers are grown in plastic greenhouses, some of which are heated (especially for roses). Kenya's climate is so mild that a wider variety of crops can be grown without cover, including carnations, spray roses and statice.

Local climatic conditions, as well as cost and ultimate quality, affect decisions about whether or not to produce crops under cover. In general, a higher degree of protection and climate control is required for higher value crops, such as roses and propagation material, and for most crops in harsher climates. For example, to produce a certain crop Dutch growers may need heated glasshouses with supplementary lighting, artificial growing media, CO_2 enhancement and water recirculation, while growers in warmer regions may need only unheated plastic greenhouses, shade houses or no cover at all. Even in relatively mild climates, however, higher yields and consistently high quality production may require investment in protective structures. Growers in regions which formerly grew crops in the open, such as Israel's Jordan Valley (chrysanthemums), and who were successful when markets were less competitive, are now considering covered production in heated plastic greenhouses.

In any event, the same crops grown with more, less or no protective cover compete with each other in the same market segments, where they are subject to similar marketing demands and consumption trends regardless of the technology used in their production. For this reason, and for our purposes, all fresh vegetables and all fresh cut flowers will be considered "protected crops".

PERISHABILITY

A critical factor in the marketing of fresh salad vegetables and cut flowers is their high degree of perishability, i.e., their extremely short shelf-life or vase-life. Most of these products have a post-harvest life ranging from only a few days (flowers such as irises and gladiolas) to one to two weeks (roses, vine-ripe tomatoes and cucumbers), three to four weeks (carnations and chrysanthemums), and even four to ten weeks (tomatoes, cucumbers and peppers when stored in hypobaric low-pressure conditions) (ITC, Burg). Spoilage of all products is much greater and faster if they are stored and handled in sub-optimal conditions of temperature, humidity and air pressure, which differ for each product. Once harvested, vegetables and flowers continue to ripen; this process can be delayed only slightly. Beyond a certain stage of ripeness, the quality and usability of these products deteriorates rapidly. Disastrous quality losses can occur at any stage in the marketing chain from grower to consumer, and the total value of the product may be lost.

In order to increase the post-harvest life of vegetables and flowers, many products are harvested before they ripen (green tomatoes or flowers with closed buds). The grower takes a calculated risk that the flower or vegetable will ripen satisfactorily further along in the marketing chain, sometimes with assistance. While this is convenient for wholesalers and retailers, the result is that consumers often buy products which never properly ripen and thus never supply the intended and expected flavor, aroma, texture and overall satisfaction.

Furthermore, since the stage of ripeness and product quality constantly changes, tomatoes picked one day with uniform green color may not be uniformly red a few days later. Such differential rates of ripening in perishable produce has given rise to the industry of re-sorting and re-packing the product in the middle of the marketing chain. A similar function is performed for cut flowers in part by bouquet-makers and floral designers who re-sort the product and discard the flowers which are unusable at that point in the marketing chain.

Because of the perishability of fresh produce, wholesalers and especially retailers have a high percentage of "shrink"—produce which deteriorates beyond marketability and must be discarded at a loss. Although retailers have always sought to minimize shrink, some industry professionals believe that retailers' shrink should be even greater than is generally the case, in order to ensure that low-quality produce does not reach the market and eventually disappoint and deter customers from future purchases (Prevor). One British retail chain, Marks and Spencer, has even begun printing a "sell-by" date on bunches of cut flowers to guarantee a post sale vase-life of at least one week.

Produce shelf-life depends on transportation, projected sales date, and intended use: vegetables and flowers produced for local sale can be harvested at a riper stage than export products because they require less transport and handling time, and veg-

etables and flowers grown under contract with retailers can also be harvested at a more advanced ripening stage because they do not need extra time to find buyers. Many uses of cut flowers, such as for weddings and funerals, require ripe flowers; since, however, the selection of special occasion flowers is made by the retail florist, this use does not affect the grower's decision about the proper stage of ripeness for harvesting.

Given the highly perishable nature of fresh produce and especially of cut flowers, post-harvest treatment and handling become critical to maximizing the preservation of product quality. Preservation of cut flowers may include immersion of flowers in water immediately after harvest, careful sorting and packing, pre-cooling, cold transport and storage, keeping the flowers away from ethylene-producing produce, adding floral preservatives to vase water and renewing the cut of the flower stems after shipment.

Every activity in the production and marketing chain must be precisely timed, especially if there is a specific target sales date such as a major holiday, as is often the case with cut flowers. For both fresh vegetables and cut flowers, there is little room for error or delay in the delivery of production inputs and the planting of propagation material and, at a later stage, in harvesting, shipping and handling the final product.

One aspect of the trade in highly perishable fresh vegetables and cut flowers which differs from the trade in storable commodities is that large surplus stocks cannot accumulate to be carried over from one sales period to another. The Dutch flower auctions are determined to prevent such a disruptive situation on even a short-term basis: a strict policy forbids carrying over unsold flowers from even one day to the next, and all unsold products are destroyed and therefore removed from the market.

While the perishability of fresh produce helps to naturally clear the market of surpluses, it also affects the power relationships between sellers (especially growers) and buyers. Transaction speed is critically important, even at the expense of downward price risk. Growers are placed under great pressure: they and other sellers in the marketing chain cannot afford to hold the product long and wait for higher prices. This situation led to the establishment of the Dutch auction method for selling perishable fresh vegetables, fruit, cut flowers and ornamental plants. This auction method gives growers greater countervailing power against the buyers by forcing buyers to compete against each other in public.

The perishability of fresh produce also makes exporters vulnerable to erroneous quality claims by import agents. The abuse of these claims can be devastating, especially when dealing in cut flowers before key holiday periods, the time when sellers expect unusually high prices. Valuable time cannot be lost in resolving such disputes when they occur. It is, of course, always conceivable that fresh produce can deteriorate in quality en route from the exporter to the importer, but this message is often received with disbelief by the exporter. For this reason, it is crucial that trading partners establish a personal relationship of trust and honesty and a method for checking claims beforehand in order to avoid problems and facilitate the smooth flow of long-distance sales.[3]

Government authorities can also exploit the perishable nature of fresh produce for protectionist purposes. This occurs when customs officials or plant health inspectors cause unreasonable delays at ports of entry. One attempt to prevent such delays and minimize uncertainty regarding the acceptability of high-value long-distance shipments has been the stationing of Japanese plant health inspectors in the Dutch flower auctions (at the expense of the auctions) to pre-inspect shipments to Japan.

High perishability also makes efficient freight transportation a critical factor for long-distance suppliers of fresh produce. Higher value products often justify the higher cost modes of transport when timely delivery is paramount. For instance, many Moroccan exporters of winter tomatoes ship their produce via trucks rather than by sea, even though truck shipments are four times more expensive, because truck transport affords them greater flexibility and punctuality (Fruit and Vegetable Markets). In other cases, exporters ship products by costly airfreight to reach the market before competitors shipping by sea.

For intra-continental supplies, exporters need access to a modern highway network and efficient refrigerated trucking services. For intercontinental supplies, exporters require sufficient air-freight capacity and regular, reliable, and reasonably priced flights. For this reason, landlocked countries such as Zimbabwe are not at a great disadvantage if their production area is near a well-served airport. However, countries which serve as regional air-transport hubs, such as Kenya, do have a considerable advantage because of access to multiple flights. Growers in countries with more inbound flights (both freight and passenger) and a critical mass of high-value perishable export products have an even greater advantage in that they may be able to act cooperatively to negotiate highly favorable freight rates and regular daily flights to major markets in the northern hemisphere. This is the situation in Israel.

SEASONALITY

Seasonality, as well as perishability, plays a critical role in marketing fresh produce. Seasonality includes not only the non-storability of fresh horticultural products but also the variations in seasonal quality, seasonal demand and the seasonality of production.

Like other agricultural crops, horticultural products have a seasonal production cycle. But unlike storable commodities, fresh vegetables and cut flowers are generally available in a given location only during their natural production season. This production season can be extended by breeding new varieties and by using horticultural techniques, including protective covering. The period of availability can be extended by improved storage methods and by importing the product from other regions having different natural seasons (Ritson & Swinbank).

Seasonality in horticultural production is the result of differences in climatic conditions, which are more pronounced in higher latitudes. Global (solar) radiation is a key climatic factor in horticultural crop production, a factor which is subject to extreme seasonal fluctuations in northern countries such as the Netherlands. Table 1 below presents the 30-year (1950 to 1980) average global radiation (in kJoule per cm^2) per four-week period in the Netherlands:

Table 1. Seasonality of Global (Solar) Radiation in the Netherlands (average 1950 to 1980)

Period	Months	kJoule/cm^2	% of Annual Total
1	January	6.3	1.7
2	Jan-Feb	12.7	3.4
3	Feb-Mar	21.3	5.8
4	Mar-Apr	36.1	10.0
5	Apr-May	47.1	12.7
6	May-Jun	53.9	14.6
7	Jun-July	54.1	14.6
8	July-Aug	46.9	12.7
9	Aug-Sep	39.7	10.7
10	Sep-Oct	25.5	6.9
11	Oct-Nov	14.2	3.8
12	Nov-Dec	7.1	1.9
13	December	4.9	1.3
Total	**Total**	**369.7**	**100.0**

Source: IKC

The Netherlands experiences a critical shortfall of natural light from October through March and especially from November through January, when solar radiation levels decrease to only 9% of the level during June-July. Horticultural techniques (such as wider spacing and supplementary lighting) cannot fully compensate for this deficiency, and thus winter production is limited.

The production seasons of vegetables and flowers in a northern European country such as the Netherlands are a combination of each product's natural season and its possible extension through horticultural techniques. The supply season for the major salad vegetables grown in heated greenhouses (tomatoes, peppers and cucumbers) begins slowly from February through April, peaks from May through September and diminishes in October and November, with virtually no production in December and January. The greatest seasonality occurs in crops grown primarily outdoors, such as asparagus. In this case, 86% of the annual production is concentrated in May and June (LEI-CBS).

Major crops of cut flowers exhibit widely varying seasonal supply patterns, based on their different natural seasons (winter/spring for some bulbflower crops and summer for some other types of flowers) and the different possibilities for extending these seasons through horticultural techniques. Dutch-grown roses are now available year-round but, due to the winter shortfall in solar radiation, the monthly production in January and February is only one-third the level of July and August, and rose quality in mid-winter is inferior to that in other seasons. In the case of roses, however, seasonality works to the growers' favor: due to a peak in demand for roses in February and the reduced winter supply, as well as to reduced demand during the summer, prices of Dutch roses actually peak in mid-winter. In comparison, some crops requiring very high light-intensity, such as gypsophila and gladiolas, can only be produced in the Netherlands from April/May through November/December. Major spring bulbflower crops such as tulips and narcissus are now produced in the Netherlands from

December through April/May. The lily, another major bulbflower, is available year-round, but winter production is greatly diminished because of lack of light needed for flowering and the seasonal availability of the propagation material (lily bulbs), which is not available for planting of some species at a time which will enable mid-winter flowering.

These conditions create the basic market opportunity for off-season supplies of fresh produce from lower latitudes and the southern hemisphere. However, flowering of certain crops may depend on such variable light levels and the occurrence of longer and shorter days in different periods during the year. Therefore, growers in equatorial countries having nearly constant day length, light intensity and mild temperatures may not be able to grow certain crops at all or in the absence of sophisticated horticultural techniques.

A degree of seasonal complementarity between northern Europe and foreign producers does exist. The Dutch supply peaks from March through September and the primary import season extends from November through May. This creates an overlap in the spring months of March through May. The year's largest total monthly supply therefore occurs in March, and results in a sharp drop in prices from February to March. As mentioned earlier, some products, such as gypsophila and carnations, are primarily imported in the winter months at a time when there is no direct competition between imports and local producers. The potential market opportunities for foreign suppliers of other products, such as roses and chrysanthemums, is curtailed by increasing Dutch production of most varieties even in mid-winter due to recent technical innovations (VBN).

The present supply pattern results from the combination of seasonal production possibilities and seasonal market opportunities in an open and competitive market. Originally, the seasonal import supply pattern for cut flowers in Europe was limited by the restrictions imposed by the Dutch auctions and by seasonal differentials in EC tariff rates (November-May and May-October). Now, however, the auctions permit some higher-quality import flowers even during the summer months, and most major non-EC suppliers of flowers pay either reduced duty rates (Israel) or no duties at all (Spain, Colombia, Kenya and Zimbabwe).

Seasonality should produce higher prices outside of the natural peak production season. Such price patterns create the economic incentive to invest in methods of extending the natural production season or to import products from regions with different natural seasons. However, EC seasonal import controls (reference prices) have distorted this price pattern for many fresh vegetables, causing market prices in the peak European production season to actually be higher than prices in the off-season. One study found only three products—melons, strawberries and courgettes— in the U.K. market which exhibit a traditional, prolonged off-season characterized by significantly higher prices and a fourth product—asparagus—with two shorter off-seasons. Fairly steady year-round supplies of all other fresh fruits and vegetables, from a combination of local production and imports, leave only very narrow "windows of opportunity" in which higher prices are sustained for periods lasting up to a few weeks (Ritson & Swinbank).

At one time, consumer demand for fresh fruits and vegetables also shifted with the seasons as different products came on the market. But now that technology can supply most products even during the traditional off-season, affluent consumers have come to expect virtually all fruits and vegetables to be available year-round (Ritson & Swinbank), although consumers may still prefer fresher locally produced products in their natural season.

Fresh cut flowers, in contrast, still display considerable seasonal supply fluctuations in most northern hemisphere markets. Imports do not fully compensate for seasonal shortfalls in local production. Some products with short vase-life, such as peonies, are not imported at all and therefore are available only during a very limited period. As a result, a very seasonal demand pattern continues for such floral products. Consumers are willing to pay very high premier prices for the first small quantities reaching the market; after this time, prices fall gradually as production reaches its peak. Prices usually rise again toward the end of the supply season as quantities diminish, although they are not likely to reach the high levels of the beginning of the season.

Another factor affecting the pricing of flowers is the fact that most floral products are still used primarily as gifts for special occasions and not for everyday personal use. A major exception is the Netherlands, where 42% of cut flowers are purchased for consumers' own use (Rabobank-F). Therefore, demand for certain flowers peaks around specific holidays and seasons, such as Christmas-New Year (mid December to early January), St. Valentine's Day (February 14), and traditional wedding seasons (late spring). Since many of these special occasions occur in the winter off-season of northern hemisphere production, prices usually reach their annual highs at these times.

There is also a traditional seasonal preference for certain colors and types of flowers in northern hemisphere markets. Oranges and golds ("fall" colors) are preferred for All Saints' Day/Halloween (October 31-November 1), reds for Christmas and St. Valentine's Day, softer colors—pink, white and especially yellow—for Easter (March/April) and Mother's Day (May in most countries). Bulb flowers are traditionally associated with the coming of spring, especially daffodils (narcissus) in the U.K. and tulips in the Netherlands. Roses, the leading cut flower in Europe, sell well throughout the year, but their romantic image makes them especially popular for St. Valentine's Day.

The United States exhibitsess strongly developed seasonal preferences than Europe, except for a few flowers. This is partly due to the fact that the U.S. has a less well-developed culture of floral consumption and partly because affluent urban Americans now expect virtually all products to be available 24-hours-a-day, year-round. Furthermore, the major supplier of the most popular cut flowers in the United States (roses, carnations and chrysanthemums) is now equatorial Colombia, which has steady year-round production and exports. Nevertheless, the U.S. still exhibits extreme seasonal demand preferences for some products such as poinsettias (for Christmas) and potted longiflorum lilies (for Easter). Some American floral industry leaders strongly believe that to develop stronger consumer demand for flowers natural seasonality should be respected, and not all products should be made available year-round (Cathey).

Peak periods of demand for cut flowers have important implications for the trade and for growers attempting to force their crop to bloom in time for these specific occasions. Wholesalers and importers begin to purchase larger quantities of flowers, especially of the less perishable varieties, about two weeks before a given holiday to allow time for re-export and distribution to their more distant customers. Heavy purchases for distribution to increasingly nearby markets, as well as higher prices, usually continue until shortly before the holiday. Auction and wholesale prices generally peak a few days before the actual holiday and then decline slightly; the fall in prices reflects the need to allow time for distribution of the product to retailers. Good wholesale demand and high prices often continue during the week after a major holiday since traders and retailers need to replenish their stocks to levels sufficient for meeting everyday non-holiday related needs.[4]

QUALITY

Quality, as well as consumer preference, is critically important in marketing fresh vegetables and cut flowers since they are intended for direct final consumption. But since these items result from a biological production process subject to variable climatic conditions and are usually handled through a long and uncoordinated marketing channel, many of their key quality attributes may be variable and unstable. A further source of complexity is that much of the appeal of fresh produce is aesthetic and emotional and, therefore, many important quality attributes of fruits, vegetables and flowers will be subjective, intangible, invisible, indescribable and immeasurable (Melamed).

The concept of quality in fresh vegetables and cut flowers embodies a multitude of diverse aspects which are perceived differently at different stages in the marketing chain. Many produce quality characteristics are difficult and sometimes even impossible to define and agree upon (Melamed). It is inevitable that products will be initially judged by their appearance and, if no other information is offered, consumers will deduce good taste and good internal quality from good, visible characteristics (Deters, Alvensleben and Meier). These include the product's color, uniformity, and freedom from blemishes, pests and diseases as evaluated against some ideal concept of that particular product/variety. Appearance quality also includes the product presentation in the sales unit (carton, bunch, etc.), and its sorting, grading and packing, which is of primary importance to the wholesaler.

Size is another key external characteristic of the product, which can be measured and valued differently in different markets. Vegetable size is mostly measured by either the diameter or circumference of the fruit, the count per carton or weight. But higher measures do not necessarily correlate with greater demand and higher prices: U.K. consumers prefer medium-size tomatoes (47-57 mm.), while German and French consumers prefer larger tomatoes (57-67 mm.) (Seker).

Flower size can be measured by the stem length, stem weight or the number of blooms per stem. A certain stem length in cut flowers is preferred, but there are negligible differences in price between those that fit this ideal and those that do not: the various shorter stems are all too short for the intended use and the longer lengths must all be cut down to the desired length. Some longer lengths may receive a small premium, but others may be sold for less; longer lengths require work—cutting—by

the buyer, and therefore shorter stems close to the desire length (about 50 cm.) may actually receive a higher price tha longer length stems (about 55 or 60 cm.) (Batt, IFB).

Quality can also be a seasonal concept, with "good" qualit being defined according to the point in the marketing chain an the time of the year. Because vegetables and flowers mature a varying rates depending on climatic conditions, they must b harvested riper, or rawer, at different times of the year in orde to ultimately attain the same degree of ripeness by the tim they reach the final consumer. For example, the minimur color standards for blocks (lots) of tomatoes in the Dutch veg etable auctions change throughout the year. Since tomatoe ripen faster in summer than in winter, greener tomatoes ar allowed to be sold in the summer auctions because they wi ripen before consumption, but may not be sold in the winte because they will not be able to ripen sufficiently (Grower).

MARKETING IMPLICATIONS

Marketing high-quality produce begins with consistently high quality production. For fresh vegetables and cut flowers, thi often requires some degree of protective covering and climate control. Protected agriculture is increasingly necessary to mee the growing demand by affluent consumers for quality product and in order for producers to differentiate their products in the crowded and competitive mass markets. In the case of cut flow ers, Dutch market research has demonstrated that growth in quantity is outpacing price increases, and suggests that produc differentiation and the creation of new market segments is the only way to escape diminishing profitability (Blauw).

Until quite recently, the production and export of protectec crops was dominated by developed country producers, such a the Netherlands and the U.S. These countries developed covered horticultural methods and technology in order to extend their own natural production seasons and improve quality tc better supply their own domestic markets. They then begar exporting these products to exploit similar off-season and high-quality marketing opportunities in neighboring countries, such as Germany, France, U.K. and Canada. Until recently, only producers in these developed countries had the experience methodology, technology and capital to practice protected agri culture on a large scale.

Producers in developing countries seeking to export protected crops to developed country markets will, first and foremost have to compete directly with such high-technology and high-quality local production, especially if such production is available during the same season. Local products may have higher production costs, but this will be compensated by more efficient labor, cheaper inputs, research and extension advice, and reduced transport costs. Moreover, local producers can supply fresher products, are familiar with local trading practices, consumer preferences and market opportunities and can better communicate with local customers.

In fact, producers in developing countries may not necessarily enjoy large cost advantages compared to producers in developed countries, when all factors are considered. For example, it is estimated that per unit production costs of vegetables in Mexico and the U.S. are often comparable, due to lower yields in Mexico and many of the above advantages for U.S. producers (American Farm Bureau). Furthermore, produce buyers nd consumers often prefer local products over imports because of presumed greater freshness (Brooker, et al, Brumfield and Adelaja)—a critical factor in the case of perishbles. Experience in the Dutch flower auctions indicates that an mported product must be of better quality than the equivalent Dutch product in order to receive the same price, and of significantly better quality to receive a higher price (IFB).

One way for foreign suppliers to overcome such a bias owards local products is to become integrated into the marketng network of a local supplier, which gives the import product he higher-quality reputation of the local product. Israel's cut lower sector does this effectively by selling most of its products hrough the Dutch auctions. These products are then mostly reexported by the initial buyers and presented as Dutch products, which is true to the extent that they now include the value-added services provided by Dutch exporters. An ironic result of his practice was observed in the mid-1980's, when carnations from the Netherlands (most of which had been imported from Israel) received higher prices in mid-winter in Germany than carnations imported directly from Israel (Staal). However, when grower/exporters market through local northern hemisphere suppliers, most of the profits are made by importers and distributors, with only indirect benefits to grower/exporters.

Another possibility for overcoming local product bias is joint venture production and marketing with a partner from a developed country. In this case, a high-quality producer or trader in a developed country contracts with specific growers in a developing country to produce and pack products according to the former's high standards in his off-season, which he can then distribute to his established customers. One such successful example is the Moroccan-French joint venture company Maraissa, which grows tomatoes in Morocco on 100 ha. for export to Europe. The tomatoes are grown, sorted and packed according to the standards and under the strict supervision of the French partner, S.M.O. from Brest (Brittany). They are then marketed in France in the winter under the Azura brand name, filling the gap between the seasons of S.M.O.'s Saveol brand from northern France and Soproma brand from southern France (Nunnink). Similar ownership or long-term contractual links have also been an important part of the development of Kenya's off-season fruit and vegetable trade with Western Europe (Jaffee).

A long-distance supplier's reputation concerning the freshness of his product is especially problematic if he lacks access to daily transport. If, for example, a given country has only two flights per week to a certain market, an importer cannot know which flowers in the shipment have been in transit for one day and which for four days. One cannot simply calculate an average age for the produce in the shipment because the four day old flowers deteriorate before those picked more recently. In this case, the importer must relate to the whole shipment as if it were four days old and pay a reduced price for all flowers.

The complex quality factors, combined with the seasonal nature of vegetable and flower production, creates a multitude of trade opportunities, sometimes in unlikely directions. Some producing and exporting countries are now importing significant and increasing quantities of fresh vegetables and cut flowers in their local off-season and even throughout the year. This is happening in response to consumer demand in both devel-

oped and developing countries for high quality, off-season produce, a demand which often cannot be supplied by local producers. It is also helped by the reduction or elimination of many trade barriers.

Producers of protected crops face several decisions which directly affect their ability to produce high-quality products. A correlation exists between per-unit production costs and high-quality production; there can also be a trade-off between productivity and high-quality. Three major production factors are all obtained at the expense of higher quality: lower investment, larger scale, and higher yields.

As discussed earlier, higher and more consistent quality production requires greater control of the growing environment. This in turn necessitates more tightly sealed structures (higher quality plastics or glass), more precise temperature and humidity control, possibly artificial growing media and supplementary lighting and improved post-harvest facilities. The construction, operation and maintenance of such facilities adds significantly to the cost of production.

Producers also must decide on the scale of production. Investors with sufficient capital and in countries with little intra-industry cooperation will find large scale production tempting, since it permits them to achieve certain economies of scale and thus lower average per-unit production costs. However, due to the extreme sensitivity of the product, horticultural production is a very intensive and delicate process which requires the constant monitoring of minute details. It also demands expert management and supervision, and that manual tasks be executed with a personal touch. Inevitably, some degree of quality will be sacrificed in a large-scale operation.[5]

The third and perhaps most critical production-quality trade-off is that of higher yield per unit of production area versus higher product quality. If the price system does not adequately reward producers for higher quality (or penalize them for lower quality), then an incentive will be created to maximize the quantity of production per unit of area (subject to marginal costs) in order to maximize revenue and profits. This higher productivity can be achieved in horticultural production by such practices as manipulating nutrient and irrigation levels and condensing plant spacing. But these same practices will also lower the quality of the produce: the result will be watery, less meaty and less tasty tomatoes and cut flowers with weaker and lighter-weight stems. The Dutch glasshouse industry boasts of its technical achievement of producing as much as 50 kilograms of tomatoes per square meter between March and November (Ministry of Agriculture-1992), but such high yields may have come at the expense of higher quality and may explain the recent decline in sales and prices of Dutch tomatoes in key markets like France and the U.K.

Producers for mass-market retail customers can realize higher prices when they supply services which add value to their products. Dutch exporters, for instance, receive premium prices for freshest vegetables and flowers in many European markets (Tradstat)—despite the frequent weakness in product quality. This is so because the Dutch provide consistent and reliable sorting and grading, high-quality packaging, a wide assortment of products and good customer service.

One study estimates that the market for premium quality

fresh tomatoes and other vegetables is only 10% to 15% of the total market. Consequently, large growers who supply mass-market retail chains prefer to produce maximum yields of mediocre quality products, at least for the time being. Many of the same growers also believe that the markets of the future will demand only high-quality produce, and they are therefore constantly testing new varieties for better taste and reasonable productivity in order to be prepared when market conditions change (Seker). While the belief that higher quality will become a minimum requirement to compete in future produce markets is widely shared (Ritson & Swinbank), no one can predict when this might occur. In the meantime, every producer and exporter must decide about the short and long-term importance of investing in quality, and develop and manage their reputations accordingly.

INFORMATION NEEDS

Given the extreme complexity of marketing fresh vegetables and cut flowers, there is a great need for comprehensive and up-to-date information about every aspect of production and marketing. Market-oriented production requires continuous market research and a smooth flow of information in order to keep up with changing supply and demand, consumer tastes and market regulations. Producers also need methods that suit their local growing conditions, research to solve future problems and to develop and adapt new products, and an advisory service to communicate research results to growers and growers' needs to researchers.

Local research is especially important for protected crop varieties which were bred and developed in other countries and for other production conditions. First of all, product selection is often difficult because each product category may contain a few to a few hundred commercial varieties. Each rose, carnation, chrysanthemum or gerbera can have its own life cycle, as well as supply and demand trends, which can themselves differ in each major market. Second, it is difficult to predict which varieties will perform well in new and untested conditions. For example, one major European rose breeder is located in northern Germany and another in southern France with its Mediterranean climate. But in Israel's Mediterranean setting, the German varieties, and not the French, actually grow better and have become the backbone of the Israeli rose industry.

As information requirements have grown, so has the cost of generating and disseminating the needed data—with the result that information has become even less available. Governments in both developed and developing countries have drastically reduced support for agricultural and horticultural research, closed experiment stations, phased out university research programs and fully or partially eliminated public extension advisory services (in cases where these ever existed). In the early 1980's, the U.S. stopped funding the collection of floriculture statistics, a service which has since resumed. The largest flower auction in the Netherlands, the VBA in Aalsmeer, cited high publication costs in ceasing publication of its statistical yearbook, which did not, in any event, include sufficient details of most interest to growers and traders (Aalsmeer Nieuws). Furthermore, the accelerated integration of the European Community into a single economic entity, eliminating internal borders, has raised concerns about the future availability of intra-EC trade data.

In other words, while the information needs of producers of protected crops constantly increase, the broad dissemination of such information is being curtailed. Market participants are increasing their private collections of such information—an often costly endeavor. To disseminate the requisite information, producers and traders across different countries and continents must cooperate and collaborate. Governments in countries with new producers of protected crops will have to ensure the provision of adequate information about domestic markets and assist exporters in gathering the necessary data about developments in foreign markets. In the meantime, the information gathering capability of the sector organizations in Dutch horticulture will add to the already substantial comparative advantages of Dutch exporters of protected crops.

2

MARKETS AND TRADE OPPORTUNITIES FOR PROTECTED CROPS

The local domestic market is an important—and often underestimated—outlet for protected crop production in middle-income developing countries. Nearly every country has its own seasonal production limitations as well as its own developing segment of urban middle-class and upper-income consumers who may be interested in off-season, higher quality and higher priced fresh produce. If the market has a well-functioning price system, seasonal production shortfalls and unsatisfied demand for higher quality produce will create domestic market opportunities for protected crops which are at least as attractive as export possibilities.

In Mexico, the domestic market has emerged as the principal sales outlet for producers who previously concentrated on exporting off-season fresh vegetables. After joining the GATT in 1986, Mexico's population became familiar with higher quality imports just as renewed economic growth enabled a consumer spending spree. Domestic growth, combined with seasonal production shortfalls and the maturity and slow growth of Mexico's leading export market, the U.S., has resulted in prices for tomatoes and other fresh produce that are higher on occasion in Mexico than in the U.S. Such price inversions intensify the redirection of Mexican export production to the domestic market. In fact, Sinaloa—Mexico's leading horticultural export state—now sells more than half of its winter tomato production domestically compared to a decade ago, when 70% of its production was exported (American Farm Bureau). Local producers of high-quality protected crops are well-positioned to take advantage of growth in domestic demand and, if high-quality imports are also allowed in the market, may be the only domestic producers capable of satisfying higher standards and competing with such imports.

China and Turkey are among the other countries which have rapidly expanded fresh vegetable production, especially under cover, but which export negligible or decreasing quantities. Both nations are primarily engaged in satisfying their own large and growing domestic needs for year-round fresh vegetables. In both countries, however, the domestic market has generally low standards, which does not stimulate innovation or quality high enough for exports (Rabobank-V).

The domestic market is an excellent training ground for building a country's export competitiveness (Jaffee). A developed domestic market is also important in that it absorbs surplus export production and production in periods falling outside narrow export "market windows," as well as produce which does not meet export standards. Even where domestic prices are relatively low, domestic sales can still be more profitable than exports due to reduced packing and transport costs (Nunnink). In any event, a well-developed local market for protected crops can indirectly improve a country's reputation as a high-quality supplier since only the best quality produce will be exported. Carnation exports from Spain in recent years illustrate this process.

Domestic markets also involve less risk and expense than exporting, with much simpler logistics, communication, payment collection and better access to market information—a major advantage for protected crop producers. In developing countries, domestic markets are often somewhat protected from imports. Local producers thus have the opportunity to be the first suppliers of high-quality produce to a captive market and to firmly establish their connections and reputations with local customers before the future arrival of imports. On the other hand, such insulated local producers may have to carry the burden themselves of developing wholesale and consumer demand for higher quality produce.

An effective wholesale and distribution system is a prerequisite for the successful establishment of high-quality domestic fresh produce supplies. Such a system must include the establishment and enforcement of product quality standards and a price-discovery mechanism which recognizes differences in product quality as well as changes in supply and demand. Because of the intensity and complexity of managing protected crop production, the grower will not also be able to engage in trading and distribution, unless he neglects his primary responsibilities. Furthermore, growers cannot risk receiving the same price for protected crops, which require high investment, as for open-field crops. Therefore, a price system must exist which offers a premium to producers of higher quality and off-season protected crops in return for their higher investment and risk. Similarly, a market channel must be in place that facilitates the timely distribution and sale of such perishable products while maintaining their high-quality.

3

INTERNATIONAL MARKETS: NORTH AMERICA

U.S. MARKET TRENDS: VEGETABLES

The U.S. is the single largest market in the world for fresh vegetables, with sales in 1991 valued at nearly $36 billion. Since 1987, the value of fresh vegetable sales in the U.S. has grown much faster in real terms (+19%) than the increase in overall consumer spending on food (+3%), which reached $619 billion in 1991. In terms of volume, however, U.S. per capita consumption of fresh vegetables increased by 6.2% during the same period, indicating a rise in real prices. Spending on food in the U.S. has continued to decline as a proportion of total consumer expenditure, dropping to less than 16% in 1991, while the share of spending on fresh produce as a proportion of the total spending on food has increased, reaching 5.8% for fresh vegetables (i.e., less than 1% of total consumer spending) and 2.7% for fresh fruit (Euromonitor).

Protected agricultural production and greenhouse produce are much less common in the U.S. than in many other developed countries. This is due to several factors: the U.S. is a large and open internal market with a well-developed road-transport system and its own warm-weather winter production regions, such as Florida and Southern California, and U.S. regulations ban the use of many chemicals considered essential for greenhouse production. Greenhouse produce is estimated at less than one percent of total produce sales in the U.S. The per capita consumption of greenhouse-produced cucumbers, for instance, is only about 45 grams in the U.S., compared to 450 grams in Canada and 4,500 grams (4.5 kilograms) in the Netherlands. In contrast to the U.S., neither Canada nor the Netherlands has year-round supplies of inexpensive field-grown salad vegetables, and greenhouse production is more competitive with imported products due to transport costs and customs duties. As a result, the Canadian province of Ontario has the largest concentration of protected vegetable production in North America (Neff).

Greenhouse produce has been sold in U.S. supermarkets since the early 1980's and retailers report small but growing consumer interest. However, they have had to invest in education campaigns to explain the higher prices of greenhouse vegetables. Some greenhouse items are becoming increasingly popular because they fill certain niches: winter hydroponic tomatoes are viewed by U.S. consumers as the tastiest alternative when field-grown vine-ripe tomatoes are unavailable. The U.S. now imports more than $40 million in greenhouse vegetables from the Netherlands, consisting mainly of peppers (80%), tomatoes, endive, cucumbers and novelties such as white eggplants (Neff, Rabobank-V).

Fresh produce sales in the U.S. and other developed country markets have been influenced by consumers' desire for healthier diets and the convenience requirements of modern lifestyles. The fresh produce industry has responded to these concerns as well as to competition from frozen and canned vegetables by creating a whole new range of pre-cut and pre-packed fresh fruit and green salads. These, in turn, have raised issues of possible contamination in the preparation of such products, but new labelling guidelines which reassure consumers also promote the healthy aspects of pre-cut salads. Fresh produce still dominates the market for fruits and vegetables in the U.S., accounting for 83.3% of sales in 1991, versus only 9.1% for canned, 5.2% for frozen and 2.4% for dried produce (Euromonitor).

Tomatoes are the leading product in the U.S. fresh vegetable market both in terms of sales value ($3.7 billion in 1990, or 12%) and nominal sales growth (+42% from 1987 to 1990). Potatoes are the second leading product ($3.5 billion), followed by lettuce ($2.9 billion) and onions ($1.4 billion). In terms of per capita consumption of fresh vegetables, the market leader is potatoes (21.5 kilograms in 1990), followed at some distance by lettuce (12.6 kilos), onions (8.4 kilos) and tomatoes (7.0 kilos). The evolution of U.S. per capita consumption of fresh vegetables (excluding potatoes) from 1970 to 1990 is presented in Table 2. In total, U.S. fresh vegetable consumption (including potatoes) reached 86.5 kilos per capita in 1990, which was almost twice the rate in northern Europe of 50 kilos but less than half southern Europe's rate of 200 kilos (Euromonitor, Rabobank-V).

Table 2. U.S. Per Capita Consumption of Fresh Vegetables (excluding potatoes), 1970—1990 (in kilograms—farm weight)

	1970	1975	1980	1985	1990
Lettuce	10.2	10.7	11.6	10.8	12.6
Onions(1)	5.6	6.1	6.0	7.5	8.5
Tomatoes	5.5	5.5	5.8	6.8	7.0
Carrots	2.7	2.9	2.8	3.0	3.6
Celery	3.3	3.1	3.4	3.2	3.3
Sweet Corn	3.5	3.5	3.0	2.9	2.9
Broccoli	0.2	0.5	0.6	1.2	1.5
Cauliflower	0.3	0.4	0.5	0.8	1.0
Honeydews	0.4	0.5	0.6	1.0	0.9
Asparagus	0.2	0.2	0.1	0.2	0.3
Other (2)	18.1	15.5	15.7	19.3	20.3
TOTAL	**50.2**	**48.9**	**50.3**	**56.5**	**61.9**

Source: USDA, Economic Research Service.
(1) Includes fresh and processing.
(2) Artichokes, eggplant, garlic, cucumbers, bell peppers, cabbage, snap beans, cantaloupe and watermelon.

Production of fresh vegetables (excluding potatoes) in the U.S. increased in volume by 18% from 1986 to 1990, from 9.2 million metric tons in 1986 to more than 10.8 million tons in 1990. Iceberg lettuce continues to be the leading vegetable crop produced in the U.S., and is of far greater importance in the U.S. than in any other country. From 1986 to 1990, lettuce production increased by 26%, from 2.6 million tons (29% of total vegetable production) to 3.3 million tons (31% of total production). Onions were the second largest crop during this period, increasing 21% to 2.4 million tons. In contrast to these two products and to its own increase in sales value, the production of tomatoes—the third largest crop—increased only 7% since 1986, stabilizing at around 1.5 million tons (Euromonitor).

Although imports account for less than 3% of fresh vegetable sales in the U.S., in 1992 they comprised 1.6 million tons valued at about $730 million. The leading import product has generally been tomatoes, in most years accounting for between 20% and 35% of both the total value and total quantity of imported fresh vegetables. In 1992, the second most valuable import was onions (15%), followed by bell peppers (12%) and cucumbers (10%). Import levels of most products fluctuate widely from year to year, depending on domestic supply conditions. As a result, the largest imported crop in terms of volume in 1992 was cantaloupe (14% of the total), followed by tomatoes, cucumbers and onions (12% each), watermelon and bell peppers (6% each), squash (5%), and carrots (4%). Table 3 shows the evolution of the U.S. import volumes of the leading fresh vegetables since 1972 (Euromonitor, USDA-ERS, USDA-FAS).

Mexico, the world's second biggest exporter of fresh vegetables (Rabobank-V) and the dominant source of imports to the U.S., regularly supplies more than one million tons of fresh vegetables to the U.S., accounting for more than 80% of total U.S. imports (American Farm Bureau). This supply is concentrated during the winter/spring period, which roughly corresponds to Florida's main production period and complements California's peak summer/fall production season.

Direct Florida-Mexico competition is reduced by differences in the regional supply patterns in major U.S. markets. Florida is the leading winter supplier to cities on the East Coast and in the Midwest; Mexico is the leading winter supplier to the West Coast. During all of 1991, Florida supplied 48% of the fresh tomatoes arriving in both New York-Newark and Chicago, compared to 20% and 27% respectively in these markets from California and only 13% and 11% from Mexico. In both these markets, 14% to 19% of the fresh tomatoes arrived from other points of origin, primarily from the southeastern U.S. states of Virginia, Tennessee and South Carolina and from local in-state production, mainly in the period from June to September. In comparison, Mexico was the largest supplier of tomatoes arriving in the Los Angeles market (40%), followed by California (38%) and Florida (21%), with only 1% originating elsewhere (USDA-AMS). Moreover, tomatoes from Florida consistently receive a $2 price premium compared to tomatoes from Sinaloa in Mexico (American Farm Bureau), further indicating that tomatoes from Florida and Mexico do not compete directly in the same quality segment.

Table 3. U.S. Imports of Major Fresh Vegetables, 1972-1992 (in metric tons)

	1972	1977	1982	1987	1992
Cantaloupe	70.5	83.1	82.9	135.7	219.2
Cucumbers	77.1	114.1	106.7	213.0	196.6
Tomatoes	266.7	359.9	223.9	414.1	196.5
Onions (all)	27.9	65.5	75.3	168.7	189.7
Watermelon	72.3	79.7	107.9	139.8	96.1
Bell Peppers	29.5	55.2	76.7	89.1	88.8
Squash	17.5	30.8	47.4	69.6	85.8
Carrots	23.2	33.0	47.8	45.3	61.0
Asparagus	—	2.0	7.3	12.9	26.2
Celery	0.9	0.6	4.6	12.4	14.9
Lettuce (all)	0.6	1.7	6.6	8.3	12.3
Broccoli	—	—	0.1	10.3	9.4
Cauliflower	0.1	0.6	4.9	6.5	8.1

Source: USDA-ERS, U.S. Department of Commerce, Bureau of the Census.

The U.S. is also a large supplier of fresh vegetables to the world market, exporting nearly 1.3 million tons of fresh vegetables in 1991 valued at about $765 million. Canada is, naturally, the largest export market for the U.S., accounting for 81% of U.S. exports by volume and 76% by value. Japan was the second leading destination for U.S. fresh vegetable exports in 1991 (6% by volume, 9% by value), followed by Mexico (3.6% by volume, 2.6% by value). The leading U.S. fresh vegetable exports are lettuce, tomatoes, broccoli, onions and asparagus, which together account for over half the total export volume and value. The USDA believes that U.S. fresh vegetable exports have grown recently due to greater trade liberalization, U.S. advantages in the areas of technology, transportation and communication links, and high quality production (Porter).

U.S. MARKET TRENDS: CUT FLOWERS

The U.S. is tied with Japan as the largest single market for cut flowers in the world, with sales in 1990 of about $6 billion. Floral purchases are still considered a luxury, though, being common for only a small percentage of the population and intended largely for special occasions. Average per capita consumption of cut flowers in the U.S. is estimated to be only 14 stems per year, which is by far the lowest of any major market. In comparison, consumers purchase 40 stems of flowers each year in Japan, 50 stems in the U.K., and 150 stems in the Netherlands. U.S. consumption is dominated by the three traditional flower types—carnations, roses and chrysanthemums—and the share of "other" flower types in the U.S. market is smaller than in any other major flower consuming country (Rabobank-F).

Because of its current low per capita consumption, low rate of market penetration and infrequency of purchasing cut flowers, the U.S. is expected to increase its demand for cut flowers at a faster rate (8% annually) than other major markets (6% in Japan and 4% in western Europe). In fact, demand is expected to double between the years 1990 and 2000 to $12 billion (van Doesburg). Increased supermarket sales are expected to contribute strongly to U.S. floral consumption; such sales, which greatly multiply the number of points of sale, currently account for only about 10% of total flower purchases. Supermarkets encourage impulse purchases by offering flowers at relatively low prices to consumers who might not purposely enter a florist shop (Rabobank-F).

The U.S. has a large and diverse floriculture sector which is located in various regions to take advantage of favorable climatic conditions and/or proximity to major consumption centers. In 1992, the top 36 floriculture producing states had nearly 10,300 commercial growers of flowers and plants with a production area of 20,000 hectares (40% under cover) and wholesale revenue of almost $2.9 billion. The states with the largest production area under cover are located in the milder climatic zones, such as Florida (3,500 covered hectares—85% of which is shade cloth and temporary cover) and California (1,500 covered hectares). By comparison, Michigan, a northern state with the third largest covered production area, has only 272 covered hectares. In terms of 1992 production value, the leading floriculture producing states are California ($666 million) and Florida ($526 million), followed by New York ($230 million), Texas ($150 million), Ohio ($128 million) and Michigan ($128 million) (USDA-NASS).

Certain states dominate the production of some types of floriculture products: cut flowers (58% in California), foliage plants (56% in Florida), and cut foliage (76% in Florida). By contrast, the production of potted flowering plants and bedding plants is more evenly distributed throughout the country, with substantial production of potted flowering plants and bedding plants in populous northern states such as New York and Michigan (USDA-NASS).

The value of U.S. floriculture production has doubled in real terms since 1981. During this period major shifts have occurred in the production pattern of each major crop category: cut flowers, potted foliage plants, potted flowering plants, and bedding plants. These developments have been accelerated by the rapid growth of cut flower imports and by the protective effects of U.S. quarantine restrictions on imported potted plants and bedding plants. As a result, bedding plants are now the leading U.S. floriculture sector, with sales in 1992 reaching $1.1 billion, followed by potted flowering plants and then cut flowers. Bedding plants now account for 38.8% of the total value of U.S. floriculture production, compared to only 25.3% in 1981. Potted flowering plants, the next largest market sector, had 1992 sales of $754 million. From 1981 to 1992, the share of potted flowering plants in the total value of U.S. floriculture production grew from only 17.8% to 26.3%. During the same period, the production value share of cut flowers decreased sharply, from 24.4% to 16.0%, for a 1992 value of $458 million. The sharpest decline in production value share within U.S. floriculture has been experienced by foliage plants, which decreased its production value 32.2% in 1981 to only 14.9% in 1992, with a wholesale value of $427 million. Cut foliage production value, on the other hand, has maintained a steady share of between 4% and 4.5% since the mid-1980's (USDA-NASS, Smith).

The U.S. is also a major exporter of floriculture products. With sales in 1991 of over $190 million, the U.S. was the fourth largest overall exporter of floriculture products, following the Netherlands ($3 billion), Denmark ($458 million), and Colombia ($332 million). The U.S. was the single largest

exporter of cut foliage ($92 million), the fifth largest exporter of plants ($84 million), and the 15th-ranked exporter of cut flowers ($14 million) (Pertwee).

Since the early 1970's, the U.S. has developed into the world's second largest importer of cut flowers (after Germany) with annual imports in excess of $300 million. Table 4 illustrates the development of U.S. import quantities of major cut flowers from 1982 to 1992. Import growth was especially rapid in the early 1980's, as the U.S. dollar increased in value relative to other currencies, reaching a peak in 1985. Since 1985, the dollar fell relative to European currencies and U.S. imports of most flowers grew more slowly, stabilized, or in some cases even declined, especially for flowers from the Netherlands and Israel (Rabobank-F, F-S Market News, Smith).

Table 4. U.S. Imports of Major Cut Flowers, 1982—1992 (in million stems, bunches (bu) or blooms (bl))

	1982	1985	1988	1992
Standard Carnations	497.6	715.4	934.5	1,154.8
Pompon Chrysanthemums (bu)	53.0	68.6	84.8	93.8
Roses	90.1	173.2	288.8	574.6
Spray Carnations (bu)	9.5	8.5	23.5	32.0
Chamaedorea (bu)	16.5	17.8		
Gypsophila (bu)	2.2	5.2	14.4	
Alstromeria	33.1	81.6	92.3	
Statice (bu)	3.4	6.5	7.6	
Tulips	13.3	59.0	41.9	61.6
Lilies	20.1	34.6	32.6	36.4
Gerbera	19.0	33.1	34.8	
Standard Chrysanthemums	26.2	40.1	26.4	34.2
Freesia	10.4	35.7	33.1	26.1
Miscellaneous Greens	74.6	14.7	6.7	24.4
Other Orchids	4.2	12.0	14.1	23.1
Iris	9.9	30.1	26.5	22.2
Leatherleaf	0.7	0.1	0.9	19.1
Gladiolus	0.8	3.9	3.2	5.1
Cymbidium Orchids (bl)	1.2	2.7	3.4	4.9
Daisies	36.4	16.7	20.2	4.8

Source: USDA, F-S Market News.

More than 70% of U.S. cut flower imports are supplied by Colombia which, since the 1970's, has developed into the world's second largest exporter of cut flowers (after the Netherlands). The import supply pattern and the effect of cut flower imports on U.S. production have varied according to type. While imports of pompon chrysanthemums increased from just 2 million bunches (of 10 stems each) in 1971 to nearly 94 million bunches in 1992 (88% from Colombia), domestic U.S. production was stable from 1971 to 1980 at about 35 million bunches, and then declined by 56% to only 15.4 million bunches in 1992. Consequently, the market share of domestic production decreased between 1971 to 1992, from 94.4% in 1971, to 47.6% in 1980, to only 14.1% in 1992.

A slightly different pattern developed for roses: although import growth was equally dramatic, domestic production actually managed to increase (but at a slower rate than imports). While rose imports grew from only 1 million stems in 1971 to nearly 575 million stems in 1992 (69% from Colombia, 15% from Ecuador), domestic production was stable from 1971 to 1980 at about 428 million stems and then increased by 25% to 534 million stems in 1992. As a result, the market share of domestic production decreased only slightly from 99.8% in 1971 to 90.6% in 1980 and then more sharply (but much less than the fall in chrysanthemum share) to 48.2% in 1992. However, U.S. imports of roses have been growing especially fast in recent years, nearly doubling in quantity between 1988 and 1992 (F-S Market News, Smith).

Colombia also dominates U.S. imports of standard carnations (97%) and spray carnations (88%), but other Latin American countries have developed as important suppliers of other types of cut flowers to the U.S. Ecuador, where Colombian producers are currently making substantial investments, is now the leading source of U.S. gypsophila imports (44% versus 39% from Colombia) and the second leading source of rose imports (15%). Costa Rica is the leading supplier to the U.S. of imported cut leatherleaf ferns (93%) and the second largest source of pompon chrysanthemums (11%). The Netherlands remains the dominant supplier to the U.S. of cut bulbflowers such as irises (99%), tulips (97%) and lilies (82%) (F-S Market News).

U.S. POLICY ENVIRONMENT

The recently negotiated North American Free Trade Agreement (NAFTA) between the U.S., Canada and Mexico, which takes effect in 1994, could affect trade in protected crops. Under NAFTA, agricultural trade agreements are bilateral. Since a U.S.-Canada Free Trade Agreement was previously established in 1988, NAFTA will have separate rules for U.S.-Mexican trade and Canadian-Mexican trade. NAFTA will generally eliminate all tariff and non-tariff barriers to horticultural trade between the U.S. and Mexico, either immediately in 1994 or during a phase-out period not to exceed 10 years for most products and 15 years for certain "highly sensitive" products (Thompson-B).

Mexican tomatoes are one major product considered import-sensitive by the U.S. and subject to safeguard provisions and a 10-year phaseout in the form of a tariff rate quota. The U.S. will charge a preferential NAFTA tariff on Mexican tomatoes up to a specified quota amount (172,300 tons from November 15 to the end of February and 165,500 tons from March 1 to May 14) and the lowest most-favored nation tariff (currently 4.6 cents per kilogram from November 15 to the end of February and 3.3 cents per kilogram from March 1 to May 14) on amounts exceeding the quota (Thompson-B). Since the current tariff rate is generally equivalent to well under 10% of the value of fresh tomatoes and since NAFTA will only reduce or eliminate this already low tariff, it is not expected to have a major impact on Mexican-U.S. trade patterns and market shares. Furthermore, specific horticultural crops in certain parts of the U.S., such as asparagus and frozen vegetables, already face stiff competition from Mexico with the current tariffs. Mexican production areas currently face such severe problems concerning other crops that the mere removal of current tariffs will not be enough to transform them into significant export products (American Farm Bureau).

NAFTA will not change the U.S. requirement that fresh produce imports from Mexico comply with U.S. standards for minimum grade, weight, size and quality (Thompson-B).

Therefore, imported tomatoes will still have to abide by the provisions of Florida's federal marketing order for tomatoes, which is in effect from October 10 to June 30. California has its own state marketing order for fresh tomatoes, administered by the California Tomato Board, which already covers Mexican tomatoes grown in Baja California and handled across the U.S. border in California. Baja producers, many of whom are joint venture partners with Southern California growers, have representatives on the California Tomato Board and participate in allocating the Board's research and market development funds (American Farm Bureau).

Other U.S. government policies exert conflicting influences on fresh produce sales. The National Cancer Institute and the California Department of Health Services began a large promotional campaign called the "Five-A-Day Program for Better Health", with the ambitious goal of inducing Americans to double their daily servings of fresh fruits and vegetables from 2.5 to five. Produce industry organizations are now trying to extend the initiative through research and education. However, for per capita demand to double by the year 2000, it must increase by 7% per year, whereas actual per capita consumption growth over the last 20 years (1970 to 1990) averaged only 1.3% per year. This development since 1970 (and particularly since 1980) has been due to increasing consumer incomes and heightened awareness of the health benefits from eating fresh produce. Even if these trends were to continue at the same rate from 1990 to 2000, education and promotional activities would have to generate an additional and unprecedented 5.7% annual increase if U.S. consumption is to double within 10 years—a prospect which is considered highly unlikely (Love). Meanwhile, the U.S. Environmental Protection Agency and the Food and Drug Administration have introduced new food safety measures which may constrain future produce sales (Euromonitor).

U.S. imports of fresh cut flowers are relatively unrestricted, with most products from most suppliers subject to a 8% tariff. Cut flowers from many developing countries, including Mexico, enter the U.S. duty-free under the Generalized System of Preferences scheme, which was renewed in July 1993. The Andean Pact countries were recently granted special duty-free access to the U.S. market in order to encourage their farmers to find alternatives to narcotics production. Special access was denied to Colombian carnations, chrysanthemums and roses on the basis that they do not need such trade concessions (Miller).

Possible non-tariff barriers to U.S. floral imports include trade legislation which permits the U.S. International Trade Commission to impose countervailing duties against foreign suppliers judged to be subsidizing exports and selling products at "less-than-fair-value" ("dumping") in the U.S. Countries penalized in recent years have included Colombia, the Netherlands and Israel. However, the main factors accounting for the heavy U.S.-orientation of Colombian floral exports are not preferential access or unfair trade practices, but rather Colombia's proximity to the U.S. market and its close ties with floral importers, distributors and retailers there. Similarly, the primary factors accounting for the decline in Dutch and Israeli floral exports to the U.S. have not been American trade sanctions, but rather the high cost of trans-Atlantic airfreight and the dramatic realignment of exchange rates since 1985 which weakened the purchasing power of the U.S. dollar relative to European currencies.

4

INTERNATIONAL MARKETS: EUROPE

Western European countries together form the largest single potential market in the northern hemisphere with over 350 million consumers, a market that is nearly equal to the United States (about 260 million) and Japan (about 125 million) combined. However, Europe is divided into different trading blocs—the European Community (EC) and the European Free Trade Area (EFTA)—which are further subdivided into 12 and 6 countries respectively. Plans call for some countries to shift from EFTA to the EC in the coming years. Significant differences also exist in the production and consumption of fresh produce from country to country and especially between northern and southern Europe. While the EC countries have accelerated the pace of their economic integration, in many respects they remain a conglomeration of distinct national markets. Since many types of production and market data have not yet been harmonized for all EC countries, this section will concentrate on some general European trends and some specific developments in the major importing countries.

Eastern European countries, especially Poland, the Czech Republic and Hungary, are having an increasing affect on the European produce market, both as consumers and suppliers. However, Eastern Europe is beyond the scope of this section.

EUROPEAN MARKET TRENDS: VEGETABLES

The fragmented nature of the European produce market creates many methodological difficulties in attempting any general analysis. Because data must be aggregated from diverse national sources covering many countries and products, no two studies present the same figures. Estimates of market value are particularly tricky, given the different exchange rates between national currencies, their fluctuations over time and different national rates of value-added tax (VAT). The EC reports statistics in European Currency Units (ECU), but these are rarely used in specific market analyses because the ECU is not yet actually employed as currency in trade. Per capita consumption data is difficult to compare across countries and over time due to the changing composition of products being measured. Nevertheless, some general conclusions can be drawn about the European market trends for vegetables.

Europe is estimated to be a larger consumer of fresh produce (all fruits and vegetables) than the U.S., with the combined market in the five largest EC countries (Germany, Italy, France, the U.K. and Spain) valued in 1991 at about $83 billion. The comparable market value in the U.S. was about $52.5 billion, with a similar number of consumers—about 260 to 280 million—in the U.S. and in these five EC countries. Italy is the largest single European produce consumer in terms of value ($29 billion), followed by France and Germany (about $17 billion each), and the U.K. and Spain (about $10 billion each) (Euromonitor). In terms of the volume of vegetable consump-

tion in the EC in 1990, Italy was again the leader with 10.4 million tons, followed by Spain (7.5 million tons), France (5.2 million tons), the U.K. (3.7 million tons), West Germany (3.2 million tons) and the Netherlands (1.3 million tons). Although total EC vegetable consumption was larger than in the U.S., it increased by only 7.6% from 1985 to 1990 compared to a 13.3% increase in the U.S. (Rabobank-V).

European consumption of fresh vegetables per capita ranges from about 50 to 60 kilograms per year in Germany and other northern European countries to about 200 kilograms in southern European countries such as Greece and Italy (Rabobank-V). Once the distorting effects of potato consumption are removed from the analysis, the different national levels of vegetable consumption are shown to be even greater. One study shows per capita consumption of fresh vegetables (excluding potatoes) ranging from a high of 65 kilograms in Spain and France to 36 kilograms in the U.K. and to only 7 kilograms in Germany (Euromonitor). Another study shows that while overall per capita consumption of fresh produce in the EC is more or less static, changes have been occurring among different countries and within the product range (Ritson and Swinbank).

Table 5 presents per capita consumption data for all fresh vegetables and for tomatoes in major EC countries from 1968/69 to 1989/90:

Table 5. Per Capita Consumption of Fresh Vegetables in the EC (in kilograms)

All Vegetables	68/69	73/74	80/81	84/85	89/90
Germany	59	67	69	72	82
France	129	113	117	118	124
Italy	165	152	165	174	167
Neth.	80	83	86	91	97
Bel-Lux	85	86	60	85	90
U.K.	61	70	78	85	65
Greece	—	215	217	194	225
Spain	—	—	160	150	223
Tomatoes					
Germany	9	13	13	14	15
France	15	15	19	21	23
Italy	36	36	41	41	43
Neth.	8	13	15	16	20
Bel-Lux	18	24	15	21	25
U.K.	—	16	12	14	15
Greece	—	116	128	92	76
Spain	—	—	33	27	40

Source: Ritson and Swinbank.

In general, European countries have experienced developments in the consumption of fresh produce similar to those in the U.S. While the trend toward healthier eating has heightened the demand for fresh produce, and specifically for leaf and salad vegetables, a concurrent desire for greater convenience has increased the demand for frozen and canned vegetables. In an attempt to reconcile these conflicting demands, a new industry of pre-packed salads has developed in some European countries. Meanwhile, traditional vegetables have generally been under pressure from new and more exotic varieties as well as other substitute food products. For instance, the consumption of onions (perceived as unhealthy) and potatoes (replaced by rice and pasta) has declined in most European countries. Germany is the exception to this trend, since consumers in former East Germany have greatly increased their purchases of potatoes and other vegetables along with other products which were hard to obtain before German unification. Another anomaly is the U.K., which is the only major market where root vegetables such as carrots are consumed in greater volume than tomatoes, the leading fresh vegetable product in the rest of Europe (Euromonitor).

More and more often, European consumers are demanding higher quality vegetables and new versions of familiar products such as tomatoes, bell peppers and lettuce, with better color, shape and flavor. As a result, the market segments for bulk vegetable commodities and lower quality products sold for processing have become saturated, which particularly affects European farmers growing vegetables in the open field (Rabobank-V). The bulk segments have come under further supply pressure as French and other northern European farmers become interested in open field vegetable production following the drop in prices for grains and other major Common Agricultural Policy (CAP) products (Fruit & Vegetable Markets).

European produce sales, particularly of pre-packed and prepared salads, are being boosted by the further development of multiples (supermarket chains). This trend is especially notable in France, the U.K., the Netherlands and Belgium. In contrast, in southern European countries such as Italy and Spain, produce sales are still dominated by small green-grocers and outdoor markets (Euromonitor).

Production of vegetables in the EC has increased by an average of 1.7% per year from 1980 to 1991, or by about 25% over the whole period. This has been achieved by improving productivity, since the total vegetable growing area has remained stable at around 1.9 million hectares. The fear that vegetable production would shift to southern European countries has not materialized: cultivation has expanded in the northern states (much of it in greenhouses) and in Spain, but has diminished in Italy, Portugal and Greece. Southern Europe has, in fact, developed into a strong market for vegetables from northern and central European countries, with France, Belgium and the Netherlands exporting considerable quantities to the south (Fruit and Vegetable Markets).

Of the nearly 5.2 million tons of fresh vegetables (excluding potatoes) imported in 1992 by the five largest EC country markets—Germany, France, the U.K., the Netherlands and Italy (PGF, Fruit & Vegetable Markets), only 853,000 tons (or 16%, valued at $804 million) came from non-EC suppliers. Imports from non-EC countries have been growing slowly, increasing by 11% between 1988 (768,000 tons) and 1992. In 1992, the three largest non-EC suppliers of fresh vegetables accounted for 55% of the value of total EC vegetable imports from non-

EC countries. The leading non-EC supplier was the Canary Islands (which is now part of the EC) with $250 million, followed by Morocco with $150 million, and Kenya with $42 million (Thompson-C).

Table 6 shows the import trends for the 16 leading products in the three major importing countries of the EC, Germany, France and the U.K., from 1988 through 1992. Tomatoes are, by far, the most important import product, accounting for about 25% of the total, followed by onions, cucumbers, bell peppers, carrots, lettuces, cauliflowers and courgettes (zucchinis). Imports of most products increased during this period, and especially in 1991, due to the consumption boom following the reunification of Germany (PGF, Fruit and Vegetable Markets).

Table 6. Fresh Vegetable Imports in Major EC Countries (Germany, France and the U.K.), by leading products, 1988 to 1992

(volume in 1,000 tons)

	1988	1989	1990	1991	1992
Tomatoes	969	1019	1009	1078	1136
Onions	681	662	685	700	605
Cucumbers	380	385	426	485	496
Peppers	268	267	270	308	343
Carrots	314	303	288	338	333
Lettuce	210	243	236	249	270
Cauliflower	235	195	168	171	209
Courgettes	78	74	80	116	97
Leeks	46	55	55	64	75
Celery	58	64	59	60	68
Mushrooms	39	44	55	60	66
Endives	68	68	68	62	65
Br.Sprouts	50	47	52	46	60
Gherkins	54	46	53	74	44
Aubergines	42	41	40	44	43
Asparagus	26	30	30	34	38
Others	573	601	627	665	733
TOTAL	**4091**	**4144**	**4201**	**4554**	**4681**

Sources: PGF, Fruit & Vegetable Markets.

Re-export is an important phenomenon in the European vegetable trade. A significant proportion of the fresh vegetables imported into some EC countries is re-exported, thereby supplying exporters during the winter and other periods when local supplies are unavailable. The Netherlands is a prime example of this trend, having re-exported exactly 50% of the 522,000 tons of fresh vegetables it imported in 1992, or 261,000 tons. For some products, such as tomatoes, the re-export rate was as high as 88% (PGF).

Since tomatoes are an important part of European diets and since they are a crop with a distinct decline in "off-season" winter production, it is natural that they should be the leading import crop. And, because tomatoes are the most important vegetable crop in terms of production value in Europe, the growing rate of tomato imports is closely monitored. Rapidly increasing tomato imports to the EC has raised great concern, especially the extension of the import season both earlier and later than in past years, which thus overlaps with local

European supplies. In 1992, tomato prices and profitability decreased sharply and a major crisis developed for suppliers of tomatoes to Europe.

From November to March, the principal supplier of tomatoes to Europe has been the Canary Islands, which would curtail shipments after April in accordance with the EC reference price scheme. From April through October, the Netherlands and other local producers mostly controlled the market thereby enjoying especially high, premier prices early in their supply season. Now, however, the market is undergoing challenges from several quarters.

The Canary Islands is now meeting strong competition from Morocco. Since joining the GATT in 1987, Morocco began lowering import duties on agricultural inputs, liberalizing laws on foreign investment and encouraging the development of horticulture. As a result, Morocco has become a major third country supplier of winter tomatoes to the EC, exporting 167,000 tons in 1992/93 (compared to about 230,000 tons from the Canary Islands) (PGF, Groenten + Fruit, Nunnink-A,B). Recently, Morocco has begun exporting even earlier than in the past by moving production areas further south near Agadir. Moroccan growers have also improved productivity by growing new varieties, expanding the growing area under protective cover, and improving quality through more selective sorting and packing. The Canary Islands has faced the increased competition from Morocco by joining the EC (which removed restrictive reference prices), and extending their export season into April and May, effectively doubling their exports in these months between 1990 and 1992.

The effect of these developments in the Canary Islands and Morocco has been serious for producers in the Netherlands and other European countries. Increased competition, for a longer period, has reduced the premier prices they received in April and May—a key factor in the profitability of glasshouse tomato production. European prices for summer tomatoes have felt still further pressure from French and Belgian growers, who have greatly expanded production and export of the tastier, more aromatic tomatoes. By November, high quality and lower priced tomatoes are again available from Spain and Morocco. Mainland Spain (as distinct from the Canary Islands) is also becoming a major player in the market for European tomato supplies, having increased exports by 55% from a 1990 low-point (PGF, Nunnink-A,B, Groenten + Fruit).

As a result, Dutch auction prices for glasshouse tomatoes—the flagship product of the Dutch glasshouse produce industry—fell by 30% in 1992 as compared to 1991. Profitability, especially for early season plantings, decreased by a similar degree and, in some cases, even became negative. Due to such disappointing results in 1992, in 1993 Dutch growers reduced the planted glasshouse tomato area by 9% (and by 23% since 1989), and increased the area of cucumbers by 9% (and by 24% since 1989) and of bell peppers by +17% (the pepper area has more than doubled since 1989). For the first time in several years, the Dutch glasshouse area for cut flowers and pot plants has been expanded, partly because some vegetable growers switched to these crops in 1993. Moroccan growers have also been hurt by the low European prices, which have caused the producer prices of even well-run farms to fall below the cost of production (Fruit & Vegetable Markets-B, PGF, Nienhuis et. al.).

In an attempt to ease the crisis of surplus tomatoes in the European market, producers in the Canary Islands, Spain, France, Belgium and the Netherlands have recently established a committee, Apetom. Among the suggestions being considered by Apetom is to request Morocco to adopt Voluntary Restraint Agreements limiting Morocco's winter tomato exports; other suggestions include extending the tomato reference price system to cover the whole year (by the EC Commission) and instituting minimum prices and production quotas (by French growers). Ironically, however, artificially high tomato prices in the EC may well encourage growers in East Europe to increase their tomato production, which could then further exacerbate the oversupply problem. At this time, no major policy changes have been instituted (Fruit & Vegetable Markets-A,B, De Kreij and Oosterhout).

EUROPEAN MARKET TRENDS: CUT FLOWERS

The western European market for cut flowers is estimated to be the largest in the world, valued at $12.5 billion in 1990, or slightly larger than the combined size of the U.S. and Japanese markets. Although the European market is expected to grow at a slower annual rate—4%—than either the Japanese or U.S. markets (6% and 8% respectively), it is still expected to reach over $19 billion by the year 2000 (van Doesburg). Western Europe is the center of world flower trade, containing the world's largest importing country (Germany), the world's largest flower exporting country (the Netherlands), and the Dutch auction system, a unique infrastructure capable of handling a large proportion of the world's cut flower and ornamental plant trade with speed and efficiency.

The floriculture industry has become one of the most dynamic sectors in European agriculture, as European demand for cut flowers and production technologies have steadily developed and expanded since World War II. Recent developments, however, have somewhat shaken assumptions about this industry. Since the early 1970's, the average annual growth in turnover on the Dutch flower auctions had been increasing (but at a decreasing rate). Then in 1992, after decades of such uninterrupted growth, the sales turnover in the auctions suddenly decreased by 2.3% from 1991, while cut flower auction prices fell by 7.5% (Hendrick). These negative results came as a major shock to the Dutch and overall European flower industry, which had made large capital investments based on the assumption of continued future market expansion. Substantial devaluations of the British pound and other major European currencies relative to the German mark late in 1992 further destabilized the industry and put many Dutch flower exporter/wholesalers into financial difficulty. In the first six months of 1993, the cut flower market enjoyed a slight rebound in the Dutch auctions, up +2.2% as compared to the same period in 1992.[6]

In 1991, the cross-border trade in cut flowers in the 12 EC countries was valued at over 3.3 billion Swiss francs, 83% of which was supplied by EC countries and only 17% by imports from non-EC sources. Of the imports originating in EC countries, 93% came from the Netherlands. (Although the Netherlands is itself a significant importer and a major re-exporter, it must be noted that 90% of Dutch flower exports comes from Dutch production and only 10% from re-exports.)

The major non-EC countries exporting to the EC—Israel, Colombia, Kenya and Thailand—account for 75% of total imports from non-EC sources. All other non-EC countries combined supplied only 4.6% of total cut flower imports in the EC, with no single country having an import share greater than 0.7% (AIPH). While these import volumes may be insignificant in the overall market picture, they can be very significant to the developing countries because of the relatively substantial foreign currency earnings they represent. Furthermore, some small non-EC suppliers are significant within their small product niches, such as South Africa for proteas and Mauritius for anthuriums.

Table 7 shows the leading non-EC cut flower suppliers to the EC in 1991, ranked by value and percent share of total EC imports.

Table 7. Leading Non-EC Cut Flower Suppliers to the EC, 1991(values in million Swiss francs; share of total EC imports in percent)

Supplier	Value	% share
Israel	166.2	5.0
Colombia	132.7	4.0
Kenya	66.8	2.0
Thailand	44.7	1.4
Canary Islands	22.8	0.7
Zimbabwe	22.2	0.7
Morocco	18.6	0.6
Turkey	17.0	0.5
South Africa	13.1	0.4
Peru	9.7	0.3
Ecuador	5.8	0.2
Brazil	4.6	0.1
Mauritius	3.9	0.1
Singapore	3.5	0.1
U.S.A.	2.5	*
Ethiopia	2.4	*
Zambia	2.3	*
New Zealand	2.3	*
Ivory Coast	2.1	*
Australia	2.0	*
Other non-EC	8.8	0.3
TOTAL non-EC	**553.8**	**16.8**

Source: AIPH, ° = less than 0.1%, 1 US dollar = 1.4353 Sfr.

The EC's import volume of cut flowers from non-EC countries increased by +70% between 1988 and 1992, far exceeding the growth in EC imports of fresh vegetables from non-EC countries during the same period (+11%). In 1992, EC cut flower imports from non-EC countries reached $450 million (Thompson-C). The Netherlands and Germany were the most significant importers of flowers from non-EC countries. The Netherlands led in imports from these countries, with 34% of the EC total, despite the fact that it is only the fourth largest cut flower importer in the EC with 8% of the total import value. Germany, the next most significant importer of flowers from non-EC countries, accounted for 25% of the total. These two countries are also the only EC countries to import a wide range of flowers from many non-EC countries. In contrast, the cut flower supplies to other EC countries from non-EC sources were usually dominated by one major source and one major product. For example,

50% of the non-EC imports to the U.K. consisted of carnations from Colombia, over 65% of non-EC imports to Italy were orchids from Thailand, and almost 50% of the non-EC imports to France were roses from Morocco (AIPH).

The trends in overall cut flower imports to EC countries have varied widely for the different major cut flower types. Table 8 illustrates the development of import quantities of the leading types of cut flowers to the EC (from EC and non-EC sources together) from 1986 through 1992. Since 1988, total cut flower imports increased by 43%. More specifically, roses increased in volume by 31%, carnations (all types) by 19%, and chrysanthemums by 18%. Of the smaller volume products, imports of orchids grew by 45%, while that of gladiolas decreased by 32%. The category showing the largest increase since 1988 was all "other cut flowers", which grew by +65% (AIPH, Tradstat).

Table 8. Import Trends of Cut Flowers in the EC, 1986-1992 (by leading flower types, in millions of stems)

	1986	1988	1990	1992
Carnations	1673	2055	2083	2446
Roses	914	1143	1355	1500
Chrysanthemums	480	502	611	594
Orchids	135	106	150	154
Gladiolas	80	69	53	47
Others (1,000 tons)	122	161	204	265
TOTAL (1,000 tons)	**251**	**323**	**382**	**461**
% of others	**48%**	**50%**	**53%**	**57%**

SOURCE: AIPH, Tradstat.

Significant differences exist in the import volume and importance of the different types of cut flowers within the individual EC country markets. As shown in Table 9, Germany is by far the leading EC importer of most types of cut flowers, including roses, carnations, chrysanthemums and gladiolas. The Netherlands is the second largest importer of major types such as carnations and roses, taking hundreds of millions of stems of each. The U.K. is the EC's second largest importer of chrysanthemums and the third largest importer of carnations and roses, while Italy is by far the largest EC importer of orchids. Table 9 also indicates that Spain has now become the EC's fourth largest importer of carnations (AIPH).

Table 9. EC Cut Flower Imports, 1992, by Market and Type (in millions of stems)

Importing Country	Roses	Carns	Chrysanth	Gladiolas	Orchids
Germany	1034	913	375	37	45
Netherlands	234	753	15	2	21
U.K.	80	627	152	1	8
France	42	26	5	1	2
Italy	10	3	4	—	73
Spain	13	73	11	—	1
Bel-Lux	49	37	21	4	3
Denmark	39	14	11	1	—

SOURCE: Tradstat.

5

DISTRIBUTION

In recent years, important developments have occurred in the distribution of cut flowers in Europe which will affect the marketing decisions of new supplying countries. The Dutch auctions are an attractive option for new suppliers, offering many potential advantages but some disadvantages as well. On one hand, the auctions offer unparalleled instant exposure and access to hundreds of the leading flower wholesalers in Europe. When selling through the auctions, it is unnecessary to conduct individual negotiations about price or payment terms, and prompt payment is guaranteed. Furthermore, the larger auctions are currently developing new sales methods to better serve new and developing market segments, such as large retail chains and organic flowers.

These advantages may be mitigated, especially for new and small suppliers, by the fact that the auctions have grown very large in recent years and must constantly confront the escalating costs of their massive logistical infrastructure. The auctions have recently introduced differential commissions and charges based on the sales costs incurred by each supplier, which will work to the disadvantage of small and irregular suppliers. The auctions have strict quality standards, which some new exporters may not be able to satisfy, and they reveal publicly which suppliers have such problems. In addition, imported flowers are usually sold after Dutch flowers, to the import suppliers' disadvantage, and Dutch buyers generally prefer local products of equivalent or better quality when they are available. However, since most new and small-scale exporters do not have the resources to market their flowers directly to other European countries, they either sell through the auctions or pay a high commission to a Dutch or German import agent.

Suppliers choosing not to sell through the auctions may elect to deal directly with the increasing number of smaller wholesalers and sometimes even with larger retailers. One advantage for the exporter is that it spreads his sales and financial risk over a greater number of customers. Dealing with smaller and more specialized customers also gives the exporter relatively greater leverage and allows him to charge higher margins, e.g. 20% instead of 10% or less.

The European Community has adopted a different policy for fresh vegetables and cut flowers than for most other agricultural commodities (which are covered by the CAP), in recognition of the seasonal and perishable nature of these products. For vegetables, this policy takes the form of intervention by producer group withdrawals, rather than purchases by official EC intervention agencies, and controlling imports through seasonally variable customs duties and reference prices, instead of import levies and export subsidies (Ritson and Swinbank). The EC's policy for cut flowers is less comprehensive than that for vegetables and it affects relatively few of the major non-EC suppliers. This section examines these policies as well as the possible implications of the recently enacted European Single Market.

For fresh vegetables eligible for withdrawal, local producer groups may withdraw produce from the markets (with some compensation at a pre-determined buy-in price) and destroy it (usually), in order to support a minimum price level during peak local production periods. For example, the eligible market withdrawal period for tomatoes is June 11 through November 30. Some commodities, such as tomatoes, also have stabilization schemes which stipulate that if withdrawals in a given season exceed more than a given percent of total production (for instance, 10%), then the buying-in prices for the following season will automatically by reduced. In this way, it is hoped, the continuation of European overproduction will be discouraged (Ritson and Swinbank).

Reference prices are also set for many fresh vegetables. Reference prices determine a minimum wholesale price level for marketing imported produce. The reference prices for each crop vary throughout their period of applicability, in order to preserve higher early season prices for local producers and to ensure that non-EC suppliers leave the market when EC production begins. Tomatoes illustrate the working of variable reference prices: the reference price (in ECU per 100 kg.) ranged from 197.27 in April to 136.75 in May and as low as 41.90 from July 11 to August 31, before increasing again to 46.47 from October 1 to December 20, after which there are no reference prices until the following April (Ritson and Swinbank).

The reference price system works as follows: when the wholesale price of a non-EC supplier, net the common customs tariff (whether or not that country pays the tariff or has a concession) falls beyond a certain margin below the reference price, for a certain number of consecutive days or days in a week, and for a significant amount of that country's produce sold in the EC (usually about 30%), then the EC Commission will levy a countervailing duty of an amount which will raise that country's "entry price" to the level of the reference price. This duty will only be removed after the non-EC supplier again abides by the reference price for a minimum period of time. Some non-EC countries have been granted limited reference price concessions, such as Morocco in the case of citrus and tomatoes (Ritson and Swinbank).

Import tariffs are another policy instrument used to control imports to the EC. As with reference prices, tariffs are generally set lower in the off-season (usually winter) and higher in the local peak production season (usually summer). For instance, the common customs tariff for tomatoes is 11% from November 1 to May 14 (with a minimum 2 ECU per 100 kg. net) and 18% from May 15 to October 31 (with a minimum 3.5 ECU per 100 kg. net) (Ritson and Swinbank). Duties on cut flowers range from 17% to 24% during the same periods. However, in the case of cut flowers, most of the significant non-EC suppliers either enjoy duty-free or reduced duty access to the EC through multi-lateral or bi-lateral concessions agreements for developing and Mediterranean countries.

A major change in the European policy environment which may affect produce imports is the enactment of the Single European Market beginning in 1993. No major change is expected that will affect fresh vegetables and cut flowers, since a single European market without internal border taxes or subsidies has been in effect prior to 1993. However, the enhanced degree of economic integration will harmonize diverse national standards and include Community-wide adoption of the strictest existing standards, all of which might negatively affect imports from some non-EC countries. The impact of these changes might be felt especially on pesticide levels on produce and packaging/re-cycling regulations (Ritson and Swinbank).

Another key area which may be adversely affected by post-1992 harmonization of national policies is phytosanitary regulations, which impose more restrictive regulations on certain types of imports. A new system of plant "health passports" was adopted in June 1993, which now requires producers of plant material to be certified by EC inspectors in the producing country, rather than in the EC market or at the EC border, as previously happened. Imported products which are presently banned in one EC country may now be banned from entire ecological regions, such as all of southern Europe. In addition, certain products which were previously acceptable in some EC countries, such as citrus plants (even for ornamental purposes), are now completely banned from import (Elhanan).

One benefit of harmonization for non-EC countries has been the termination of specific national import restrictions. These included, for example, summer and autumn import bans on tomatoes in France, Belgium and Greece from a number of countries, and minimum prices for tomato imports in France in November and December. Other restrictions now eliminated were licensing requirements for importing summer tomatoes in Denmark and Ireland (Ritson and Swinbank).

A country-specific policy change that will have a far-reaching impact on horticulture, and particularly on floriculture, is the recent environmental legislation in the Netherlands. The Dutch government has set targets for substantial reductions in the use of agricultural chemicals by the mid-1990's and the year 2000. These regulations ban the use of many chemicals, mandate the reduction in the use of others and restrict the emissions of CO_2, nitrites and phosphates (Ministry of Agriculture-

1990). The effect of this policy change will be felt immediately in cut flower production: it will require crops and most flowers that are to be grown in artificial substrates (i.e., hydroponically) with water recirculation to be transferred to the new cultivation systems (some as yet undeveloped) within a very short period of time. Consequently, the Dutch have had to redirect most of their research resources to finding solutions which will satisfy these new regulations. It is assumed that such solutions will be found for most crops, but they will require huge additional investments by growers without necessarily improving productivity or quality (and may even adversely affect these goals). Partly as a result of this new situation and in light of long-term industry trends, the Dutch Agricultural Economics Institute expects 15% to 25% of Dutch flower growers to leave the sector in the near future (Kijne et. al.).

A potentially significant effect of EC policy on horticultural trade in the 1990's is the inclusion of Iberia in the EC. Originally scheduled for 1996, the process of tariff reductions and full membership for Spain and Portugal were accelerated by the EC Council with formal admittance effected in January 1993. In the meantime, Spain and Portugal benefitted from reduced rates of countervailing duties (when applied) and from a favorable inconsistency between their special "Community Offer Prices" and the official EC reference prices. Even so, a reduced reference price-type mechanism to control Iberian exports to other EC countries remained in place.[7]

Nevertheless, Spain presently exports only about 10% of its total vegetable production, and since joining the EC, wages have risen substantially. One result has been that Spain's domestic market has absorbed most of any increased production and has even begun attracting produce imports. Production costs in Spain are expected to rise and become comparable with those in other European countries, eliminating Spain's cost advantages. Since Spanish horticulture is lacking in innovations, high quality infrastructure, education, information, research and cooperation within the sector, this study does not expect Spain to play a dominant supply role in the future European fresh produce trade (Rabobank-V).

6

INTERNATIONAL MARKETS: JAPAN

As a prosperous country with a large population which has enjoyed unparalleled economic success in recent decades, Japan has also developed into a major consumer of fresh horticultural products. Despite its huge local production, exacting quality standards and reputation as an insular and highly protectionist market, Japan has also become one of the world's largest importers of horticultural products, including fresh vegetables and cut flowers. Japanese horticultural imports in 1991 reached nearly $4.8 billion, including $444 million in fresh vegetables and $243 million in fresh cut flowers and other nursery products. The leading horticultural import categories were fresh non-citrus fruit ($774 million) and fresh citrus fruit ($538 million) (Thompson-A).

The U.S. supplied 29% of the value of Japan's total horticultural imports in 1991, or nearly $1.4 billion (cif), but only 18%, or $78 million, of Japan's fresh vegetable imports. American exporters have been successful in Japan by offering consistent availability of high-quality produce and by engaging in aggressive market promotion and consumer education. They have also benefitted greatly from U.S.-Japan bilateral phytosanitary negotiations and market access agreements which have recently opened up the Japanese market to many new horticultural products. There is potential for substantial future import growth since many temperate horticultural products are still effectively blocked from entering Japan, including fresh tomatoes and other salad vegetables (Somers and Thompson).

Quality, freshness and appearance are critical to Japanese consumers, most of whom shop daily and are willing to pay premium prices for high-quality products. A strong factor in the market is the Japanese custom of gift-giving, which in recent years has included a growing proportion of fresh produce and cut flowers. As a result, although 90% of imported foods (mostly bulk items) pass through Japan's large general trading companies, items which require special care and handling, such as fresh vegetables and cut flowers, are mainly imported by smaller specialized firms (Somers and Thompson).

JAPANESE MARKET TRENDS: VEGETABLES

While Japan's consumption of fresh vegetables has been stagnant and even declined somewhat from 16.0 million tons in 1987 to 15.6 million tons in 1991, still it remains almost 50% larger than the largest single European country market, Italy. Japan's annual per capita vegetable consumption of about 128 kilograms ranks between that of southern Europe (about 200 kilos) and the U.S. (about 87 kilos). Sector shares in Japan's vegetable market have remained remarkably consistent in recent years with root vegetables dominant, accounting for about 41% of total volume, followed by leaf vegetables (about 31%) and other vegetables (28%) (Euromonitor).

Japanese consumers' tastes in vegetables are generally very conservative and consumers are still very price conscious. In recent years, however, a demand has been developing for tastier vegetables (JETRO-V). Dutch sources report that the Japanese prefer slightly smaller tomatoes of uniform size, with sweet flesh and a pastel-red color, and red, yellow and other colored sweet peppers are preferred to green (IFW). Japanese vegetable consumption also exhibits a seasonal pattern. Peak consumption occurs in the spring (March to June) and fall (September to December) (JETRO-V).

Vegetable production is the third most important sector in Japanese agriculture, following rice and livestock. Most vegetables in Japan are grown in the open on small farms averaging only 1.5 hectares. Some salad vegetables for fresh consumption, such as tomatoes, cucumbers, aubergines and peppers, are grown under cover in glasshouses or plastic tunnels (Rabobank-V). The total vegetable growing area decreased slightly from 557,000 hectares in 1984 to 538,000 hectares in 1989. A more pronounced decline was evidenced in root vegetables and heavy vegetables such as Chinese cabbage, due to labor shortages and the aging farm population. The production volume of some salad vegetables also decreased during this period, including cucumbers (from 1.033 million tons to 975,000 tons) and tomatoes (from 802,000 tons to 773,000 tons) (JETRO-V).

Japan did not import fresh vegetables until 1969 and, although fresh vegetable imports have nearly doubled in volume from 124,000 tons in 1985 to 236,000 tons in 1990, they remain only a very small proportion of total consumption. Most Japanese imports consist of vegetables to fill off-season supply shortages (e.g., asparagus) and vegetables which either cannot be grown in Japan or for which domestic production is declining. Some vegetables are now imported year-round (carrots and broccoli), while others are supplied on a spot-basis, depending on fluctuations in market prices (JETRO-V).

The development of Japan's vegetable imports since 1985 is shown in Table 10. The major vegetables imported by Japan are generally less perishable and more durable, such as pumpkins (99,000 tons in 1990) and onions (87,000 tons in 1990), both of which can withstand long-distance transport and fumigation. Their import volumes have increased substantially since 1985 (pumpkins by 150% and onions by 40%). In contrast, the next most significant import products are much smaller in quantity: asparagus—11,600 tons in 1990, up from 2,400 tons in 1985, and broccoli—nearly 11,000 tons in 1990, up from only 383 tons in 1985. Recent technological advances in maintaining the freshness of leafy vegetables have led to increased imports of such products as lettuce, but the volume of these imports remains relatively small (JETRO-V).

TABLE 10. JAPAN'S IMPORTS OF MAJOR VEGETABLES, 1985—1990 (VOLUME IN 1,000 TONS)

	1985	1986	1987	1988	1989	1990
Pumpkins	40.9	46.1	58.4	82.0	81.8	99.2
Onions	61.2	53.0	35.5	112.4	80.8	86.6
Asparagus	2.4	5.3	8.8	11.9	10.7	11.6
Broccoli	0.4	1.4	0.3	2.2	5.4	11.0
Garlic Shoots	2.8	4.5	5.4	6.0	6.2	5.8
Garden Peas	3.7	5.7	2.5	1.7	4.9	4.0
Carrots	0.1	8.9	2.2	0.8	1.1	3.3
Green Soybeans	2.6	3.1	3.2	3.0	1.2	2.1
Okra	0.5	0.4	0.5	0.8	1.4	1.9
Cabbage	0.3	7.6	1.3	3.4	2.5	1.7
Bamboo Shoots	0.5	0.4	1.0	1.2	0.9	1.4
Cauliflower	0.0	0.0	0.0	0.8	0.7	1.3
Ginger	0.3	0.1	4.8	1.8	0.3	1.3
Total	**124**	**142**	**126**	**241**	**206**	**236**

Source: Japan Fresh Produce Import Facilitation Association.

Japanese importers have become active in developing overseas supplies by providing foreign growers with seeds of the varieties they handle. In addition, such importers provide technical advice regarding growing methods and post-harvest handling. The most significant example of such Japanese import development has been the case of pumpkins produced in New Zealand and Mexico (JETRO-V).

After years of being barred from entering the Japanese market by stringent phytosanitary restrictions, the Netherlands reached new agreements with Japan in February 1993 and began exporting tomatoes and peppers in March of that year. The Dutch plan to export to Japan fruit and vegetables valued at $15 million to $25 million by the end of 1993. The Dutch will sell only high-value products since all products must be shipped by airfreight (IFW).

In order to convince the Japanese plant health authorities that Dutch vegetables would not contain Mediterranean fruit flies—despite the Dutch claim that these insects cannot survive in the Dutch climate—Westland growers have installed fruit fly lures in their greenhouses. These lures are inspected regularly by Dutch quality-control experts together with Japanese inspectors. Products are also sampled at regular intervals and then isolated in subtropical conditions to check for pests. Tomatoes and peppers destined for Japan are also supplied in special fine-mesh packaging. Although such measures are costly, the Dutch feel these investments are necessary to break into the Japanese market and ensure long-term success (IFW).

JAPANESE MARKET TRENDS: CUT FLOWERS

Japan is tied with the U.S. as the single largest country market for cut flowers, valued at about $6 billion in 1990. However, since Japan has only about half the population of the U.S., the per capita consumption value of flowers in Japan is about twice that of the U.S. The total Japanese market for cut flowers is expected to grow at 6% per year in the 1990's compared to 8%

expected annual growth in the U.S. At these rates, the Japanese cut flower market will be valued at about $10.4 billion by the year 2000, which will place it second behind the U.S. at $12 billion (van Doesburg).

Demand for cut flowers in Japan has increased substantially in the past decade, stimulating the doubling of domestic production and the development of a significant volume of imports. Since the mid-1980's, imports of flowers to Japan have nearly tripled in volume (measured in either tons or stems) and in value (yen) from 1986 to 1991. Imports reached almost 360 million stems in 1990 (the latest figures for stems available) and 15,600 tons and 19.2 billion yen in 1991, accounting for an estimated 7.5% of Japan's total supply of cut flowers (JETRO-F).

The Japanese External Trade Resource Organization (JETRO) expects Japanese consumers to continue to spend more on flowers "as long as the Japanese economy continues to grow steadily in the coming years ..." (JETRO-F, Rabobank-F). However, by mid-1993 Japanese economic growth had slowed to the lowest rate in years. Despite the fact that economic indicators in Japan were still positive when compared to the U.S. and Europe, the Japanese became increasingly concerned about their immediate economic future. Perhaps as a result, cut flower imports declined in 1992 by 24% in volume to less than 12,000 tons and 16% in yen value to 16.1 billion yen, following a decade of steady increases (Vakblad-1993). Part of the decrease in imports may also be attributable to the augmented production in Japan of many types of cut flowers which were previously imported, such as lilies and tulips. Japan began importing considerable quantities of flowerbulbs and other propagation material from the Netherlands in 1987. In 1992, Japanese flowerbulb imports from the Netherlands increased by 35% in volume from 1991, and accounted for nearly 10% of the value (about $60 million) of all Dutch flowerbulb exports (PVS, IFW, Hendrick).

In terms of volume, Taiwan is now the leading source of floral imports (5,000 tons, up in 1991 from 1,400 tons in 1986), surpassing the long-standing leader, Thailand (4,300 tons, up from 2,000 tons in 1986). The volume from the third leading supplier, the Netherlands, grew eight-fold from only 335 tons in 1986 to 2,500 tons in 1991; Australia was the fourth largest supplier (1,100 tons in 1991, a twelve-fold increase from only 84 tons in 1986).

In terms of value, the Netherlands surpassed Thailand in 1989, becoming the leading import source and reaching some 6.5 billion yen in 1991. Thailand has remained second in value of imports, with 4.2 billion yen, followed by Taiwan (1.8 billion yen), New Zealand (1.7 billion yen), Singapore (1.5 billion yen), and Australia (1.3 billion yen) (JETRO-F). From 1990 to 1992, the Netherlands' share of import value decreased slightly, from 35.5% to 32.1%, as did the shares of Thailand, Taiwan and Australia. Meanwhile, the market value shares of Singapore and New Zealand have been increasing (Vakblad-1993).

Dutch cut flower exports to Japan have consisted primarily of bulbflowers, in contrast to the traditional Dutch export assortment of roses, chrysanthemums, carnations and summer flowers. In terms of f.o.b. value in 1992, the leading flowers exported by the Netherlands to Japan were lilies (44.2%), freesias (9.3%), tulips (8.2%), nerines (7.9%) and roses (7.8%) (PVS).

Japanese import statistics show that virtually every foreign supplier of cut flowers to Japan has experienced a stagnation or even a drop in export value following their initial attempt to enter the Japanese market. Then, as they established relationships with importers, learned the market requirements and gained more access through trade agreements, many suppliers began to recover and surpass their initial trade volume and some even achieved substantial success. For example, following the Dutch-Japanese agreement in 1988 (IFW), and the subsequent stationing of Japanese plant health inspectors in the Netherlands (at Dutch expense), Dutch exports of cut flowers to Japan accelerated rapidly, more than tripling in value in two years, from 1.7 billion yen in 1987 to 5.7 billion yen in 1989 (JETRO-F).

There are two channels of distribution for cut flowers in Japan: the many wholesale markets, or auctions, which account for an estimated 80% of the flower trade, and another channel which uses direct negotiations with (mainly) retail chains. In 1990, there were 17 central and 229 regional wholesale markets, with a further 80 small-scale independent wholesale markets (JETRO-F). Tokyo alone has about 30 flower wholesale markets/auctions (Egberts), compared to only nine in all of the Netherlands.

CONSUMER PREFERENCES

Japanese preferences for and uses of cut flowers differ in many respects from those in the U.S. and Europe, and these factors must be considered by potential exporters to Japan. The primary consumers of cut flowers in Japan are businesses (offices, hotels, restaurants, funerals), accounting for 40% of the market. Business-related gift-giving accounts for another 30% to 40% of the market. Private consumption accounts for the remaining 20% to 30% of Japanese demand, including 90% intended for personal gift-giving and only the small remainder used by consumers in their own homes. In other words, the self-use segment accounts for only 2% to 3% of total Japanese demand, compared to 35% in the Netherlands (JETRO-F, Egberts).

Japanese consumers prefer relatively high-quality flowers and are willing to pay high prices for them. High quality is important not only because most flower purchases in Japan are intended for gifts, but also because of the Japanese flower-arranging style which emphasizes the beauty of a single bloom or a small number of selected flowers (JETRO-F). Demand for quality creates higher retail prices since retailers must throw out a large number of past-prime flowers, sometimes accounting for as much as 40% of the flowers in a shop (Egberts).

Japanese consumer demand for cut flowers is very seasonal in nature, in accordance with the Japanese festival calendar, which differs from the U.S. and Europe. Floral demand in Japan peaks in March and September for the vernal and autumnal equinoxes, traditional times for placing flowers on ancestors' tombs. Demand also peaks in December-January for the New Year season (both Gregorian and Chinese), when it is customary to give business and personal gifts, and in June for Mother's Day, when red carnations and pink carnations and roses are especially popular (JETRO-F).

Japanese consumers have strong color preferences for flowers depending on the season, the type of flower and even the region in Japan. In general, consumers prefer subtle colors such as white, pale pink, pink, salmon pink, light purple and red. Brighter colors are gradually becoming more popular, but still account for a small share of the market. In spring, yellow is the preferred color, while summer preferences shift to cool tones such as white and blue. Purple is considered a fall color, while red is preferred in the winter. Chrysanthemums are preferred in white or pink; tulips and carnations in pink (except for Mother's Day, when red is preferred); roses in red; and statice/limonium and gentiana in purple (JETRO-F, Egberts).

DOMESTIC PRODUCTION

Flowers are produced in Japan by more than 81,000 growers in an area of 16,609 hectares, meaning that the average size of a flower farm is only about 2,000 square meters. Between 1985 and 1990, the number of flower growers increased by 10% and the acreage expanded by nearly 30% (JETRO-F), as flowers showed greater profitability than traditional crops such as rice and vegetables. During this period also, the government offered incentives to farmers to diversify away from rice production and into floriculture (Egberts). Nevertheless, the income of Japan's floriculture sector is still relatively small, measuring only 60% of the income from fruit production, 20% of the income from vegetable production and only 4.5% of total agricultural income (Ohkawa).

In 1990, Japan produced more than 5.3 billion stems of cut flowers, an increase of 18% from 1986. The major crops grown in Japan were chrysanthemums (1.9 billion stems; 36% of total production), carnations (700 million stems, or 12%) and roses (over 400 million stems, or 11%) (JETRO-F). These three crops are still expanding, but the strongest recent production growth has been in other types of flowers such as lisianthus, limonium, tulips, gladiolas and lilies. Protected production of flowers is increasing (Rabobank-F), with the majority of carnations and roses, and half the Japanese chrysanthemum crop, grown under cover (Ohkawa).

In general, Japan imports cut flowers which are either difficult to grow in Japan, unavailable during Japan's off-season, cheaper to grow overseas, or new to the Japanese market. In the coming years, the role of imports in Japan is expected to shift from filling these gaps in the market to supplying large volumes of the basic floral commodities; this could boost import shares from their present levels to 15% to 20% of the total market, which itself is expected to grow steadily. In order to reach these goals, foreign suppliers will have to establish long-term relationships with Japanese importers, who can guide them in satisfying Japan's stringent standards regarding plant health, harvesting stage, grading, packaging, labeling and consumer preferences (JETRO-F).

The development of Japan's import volumes of the major types of cut flowers (in stems) since 1984 is shown in Table 11 (JETRO-F). Japan's cut flower imports have primarily consisted of one or two specific products from each of a few main suppliers. Orchids from Thailand and Singapore, chrysanthemums from Taiwan and ferns and bear grass from the U.S. comprise two-thirds of the volume (in stems) of Japanese imports. Emerging sources of supply and more recent import products include the Netherlands (lilies and other bulbflowers), Australia (anigozanthos and waxflowers), New Zealand (zantedeschias and orchids), Costa Rica (lower-cost ferns), and Colombia (lower-cost carnations).

Table 11. Japan's Import of Major Cut Flowers, 1984—1990 (volume in million stems)

	1984	1985	1986	1987	1988	1989	1990
Orchid	49.1	63.2	78.8	84.1	115.3	112.0	124.2
Ferns	12.4	17.0	28.4	31.2	45.6	50.6	52.2
Chrysanthemum	23.3	24.6	15.9	27.7	41.2	44.1	30.5
Bear Grass	—	—	5.9	9.7	27.6	26.4	26.7
Carnation	2.9	3.3	7.0	7.6	11.5	10.1	12.9
Freesia	—	—	3.2	4.7	11.9	11.2	8.6
Anthurium	4.9	4.6	5.8	4.4	4.2	4.5	5.1
Gladiolas	0.4	1.5	3.7	3.6	6.5	9.9	6.8
Nerine	—	—	2.0	3.0	4.9	6.7	6.2
Lily	—	—	1.8	3.0	6.0	9.9	9.6
Ornithogalum	—	—	0.5	2.1	3.5	—	—
Tulip	—	—	0.9	2.2	9.4	10.6	8.2
Daisy	—	—	1.7	1.8	1.1	—	—
Ruscus	—	—	—	1.4	1.4	2.0	3.3
Waxflower	—	—	—	1.1	1.4	3.2	4.1
Anigozanthos	—	—	—	1.0	2.4	—	—
Leucadendron	—	—	0.6	0.8	0.9	0.9	1.0
Rose	—	—	—	—	—	9.9	11.0
Others	6.0	9.3	7.9	10.6	29.9	82.3	—
Total	**99**	**123**	**164**	**200**	**325**	**357**	**358**

Source: Ministry of Agriculture, Forestry and Fisheries.

JAPANESE POLICY ENVIRONMENT

The Japanese policy environment affecting the import of fresh vegetables and cut flowers consists primarily of phytosanitary import prohibition measures. Despite these restrictions, which are often viewed by exporting countries as protectionist non-tariff trade barriers, many countries have succeeded in gaining greater market access to Japan through a combination of careful compliance, patience and diplomatic pressure. Tariff barriers, on the other hand, can be significant for some crops and insignificant for others. All cut flowers, for example, enter Japan duty-free and there is only a 5% tariff on cut leaves and branches (JETRO-F).

Japanese quarantine regulations prohibit the import of a number of host plants of pests, pathogens and other materials not native to Japan and considered to pose a threat to Japanese agriculture and forestry. Some 15 species are on the current list of prohibited pests, with the Mediterranean fruit fly, the oriental fruit fly, the Queensland fruit fly and the melon fly causing the most concern. Other prohibited pests include the codling moth, the sweet potato weevil, the sweet potato vine borer, the

West Indian sweet potato weevil, the potato wart, the Colorado potato beetle, the potato cyst nematode, tobacco blue mold, the citrus burrowing nematode, the hessian fly, and other rice pests and pathogens. Entry to Japan is also prohibited for plants and materials shipped via contaminated areas, unless they are not unloaded along the way (JETRO-Q).

In addition, entry of soil and plants with soil is prohibited, except in cases where a country has developed methods of disinfection which ensure complete freedom from the prohibited pests and pathogens. The import ban may be lifted on an item-specific basis following the presentation of evidence of the disinfection method, on-site inspections by Japanese plant health experts, bilateral negotiations, public hearings and subsequent changes in the regulations (JETRO-Q).

Shipments of cut flowers found to contain prohibited pests are fumigated with either hydrocyanic gas or methyl bromide, depending on the type of pest found. In 1989, 34% of the total stems imported were fumigated, 20% of these with hydrocyanic gas and 80% with methyl bromide (JETRO-F). The latter is considered to cause significant damage to the flowers. As shipment size increases, a decreasing portion of the shipment is inspected. Japanese flower importers have recently been successful in further reducing the size of the samples required to be inspected, which will improve the chances for avoiding fumigation (Sugiyama).

The plant quarantine process can be expedited and the success rate increased by establishing a pre-clearance system whereby Japanese plant health inspectors are stationed in the exporting country. In order to justify the expense of such an arrangement, exporting countries need to have sufficient export volume to Japan, a concentration of production intended for Japan in a limited area and little history of disease and pests in cut flower shipments. After pre-clearance in the exporting country, flowers arriving in Japan require only nominal inspection at ports of entry and, as a result, can be distributed more quickly (JETRO-F).

For edible horticultural products, significant market access has been achieved through international negotiations and agreements. Under the GATT-11 agreement, Japan consented to liberalize import quotas on eleven different horticultural products between October 1988 and April 1990. With the 1988 U.S.-Japan beef/citrus agreement, Japan agreed to completely liberalize import quotas for oranges by April 1991 and for frozen concentrate orange juice by April 1992. While high tariffs (from 20% to 40%, depending on season) remain in place for oranges, imports from the U.S. have increased since the quotas were removed and are expected to increase substantially in the future (Somers and Thompson).

CONCLUSIONS - ANNEX

The production and marketing of protected crops is an infinitely complex process, which requires prospective producers to face sets of difficult choices. This paper has attempted to provide a context for some of these choices by describing many of the important factors affecting the marketing of perishable produce and by highlighting the principal trends and opportunities in the major northern hemisphere markets. After taking these factors into consideration and making the basic decision to initiate protected crop production, two key strategic issues must be resolved: selecting a primary target market (export or domestic) and deciding which general product category (vegetables or flowers) to produce with a given set of production inputs.

Many projects in protected crop production are established with the intention of developing an export industry, and are based on the assumed advantages of lower production costs and more favorable climatic conditions (especially in winter) compared to northern hemisphere developed countries. But growing consumer demand for year-round availability and higher quality is largely being met by local developed country producers and existing import sources. Therefore, opportunities for new suppliers to fill off-season supply shortages and other narrow market niches are actually diminishing and even disappearing. Consequently, protected crop production, which is considerably more complicated and expensive than open-field production, does not in itself ensure successful exports.

To date, the share of fresh produce imported from developing countries into northern hemisphere developed country markets has remained very small. Furthermore, the existing import trade is dominated by a few well-established and favorably situated exporters, such as Mexico (tomatoes to the U.S.), Colombia (flowers to the U.S.), Israel (flowers and vegetables to Europe), Kenya (flowers to Europe), and Spain, the Canary Islands and Morocco (vegetables to Europe). These countries possess combinations of many favorable elements which make it difficult for new suppliers to displace or even emulate them. Clearly, the success stories in the international produce trade are the result of much more than just low production costs, mild climate and the ability to purchase technology and know-how. A closer examination shows that the industries in most of the established produce exporting countries began developing together with the major northern hemisphere markets more than 20 or 30 years ago. In countries generally considered to be less-developed economically, these industries comprise pockets of high technological development on par with competing producers in developed countries. Over the years, these exporters have built individual reputations as reliable suppliers, gained considerable expertise and developed invaluable networks of commercial relations. Successful exporters have worked closely with partners, investors and/or importers in developed country markets, who continually help them fine-tune their production to satisfy the stringent requirements of these markets.

Consequently, many existing and prospective exporters of protected crops in developing countries may give more serious consideration to the potential of their domestic markets. In many countries, the increasing emergence of a middle-class and their growing exposure to higher-quality products creates substantial domestic market opportunities for higher quality and off-season protected crops. Local sales usually involve less transport and handling costs, fewer logistical and communication problems, and more secure payment collection than export sales. In addition, local supply and demand imbalances, especially in cases where a country has limited imports, often create a price inversion whereby local market prices are higher than export prices. These factors have led to the increasing re-orientation of horticultural producers from exports to domestic sales in countries such as Mexico and Spain.

Of course, there may also be important non-economic reasons for utilizing protected crop production technology, such as increasing the volume and reliability of food crop yields to help feed an expanding domestic population. Greater yields, conservation of limited resources, and year-round production could be achieved by using protective cover and other horticultural techniques, but would require costly investments and the establishment of new infrastructure and support services. Since the necessary technology and methodology may have to be imported at high cost, crops produced under protective cover and not sold at market prices could require subsidization.

The intended market—domestic or export—is a strong factor in determining whether to produce protected crops of vegetables or cut flowers. For prospective exporters, product choice may also depend on the producer's geographic proximity to the intended market.

For domestic sales in a developing country market, vegetables grown under protective cover may offer greater immediate promise than cut flowers. It may be easier to encourage consumers to switch to higher quality and off-season food products than to develop new consumption habits for a non-essential product such as cut flowers. Furthermore, although fresh vegetables are also perishable and require careful handling, they may be more durable and better able to withstand sub-optimal handling conditions than are fresh cut flowers.

While it will be difficult for new developing country suppliers of either cut flowers or fresh vegetables to successfully enter northern hemisphere developed country markets, cut flowers have slightly better export prospects than fresh vegetables. Developed country markets for both cut flowers and fresh vegetables are dominated by local suppliers, and the primary import suppliers are located on the nearby southern periphery of these markets. However, a significant volume of cut flower exports has developed from more distant suppliers, which has not generally been the case with fresh vegetables.

The relatively favorable outlook for cut flower exports is based on certain advantageous product characteristics, market developments and access to market channels which are not enjoyed by fresh vegetables. In general, the demand for cut flowers has been growing more rapidly in recent years than the demand for fresh vegetables, despite the fact that this rate of increase is itself diminishing through time. Furthermore, cut flowers generally have a much higher value-to-volume ratio than fresh vegetables, which is a critical factor in the profitability of products that must be transported long distances by costly airfreight. Also, since cut flowers are so perishable in nature, even producers who are close to the market must use expensive climate-controlled transport and handling, thus partially off-setting the transport cost disadvantage of long-distance suppliers.

A further disadvantage for prospective vegetable exporters to Europe is that, unlike with cut flowers, non-European suppliers cannot sell their vegetables through the Dutch auctions. Such auction sales have been instrumental in assisting countries such as Israel and Kenya in becoming become major suppliers of cut flowers to Europe. Another potential disadvantage for smaller developing countries contemplating vegetable exporting is that the leading exporters all have large domestic markets to absorb second-quality and surplus production.

A special complication for vegetable exporters is that, unlike cut flower exporters who need only be concerned about maintaining good external product appearance, vegetable suppliers must also ensure satisfactory product flavor. Consumers may have very subtle flavor preferences which differ from country to country. Accordingly, it will be easier to introduce new exotic ornamental products to the market than new exotic vegetables since, in the latter case, consumers must be willing to taste the new product and learn to like new flavors.

It is a difficult task for any producer—in either a developed or developing country—to consistently supply high-quality fresh vegetables or cut flowers. The most established and experienced producers of such sensitive biological products have production and post-harvest handling problems. These difficulties are a discouragement and an opportunity: while they necessitate complex knowledge of production and marketing, they also provide opportunities for new suppliers to try to improve on the performance of existing suppliers and better satisfy the needs of the trade.

The chances of new suppliers to enter competitive markets, find unexploited niches or displace existing suppliers may be small, but they do exist. Ultimate success or failure can only be judged over time, since each season may present special supply problems and demand developments. To successfully exploit such export market opportunities, the new producer must be prepared for a long-term investment, be willing and able to continually learn about the potential market and marketing system, and be equipped with much patience and much persistence.

ENDNOTES - ANNEX

[1] Alan J. Malter, market researcher for horticultural products in the Market Research Department of Israel's Ministry of Agriculture.

[2] Throughout this section, particular attention is given to and many examples are drawn from the Netherlands' glasshouse sector. Due to its size, concentration, institutional structure and international orientation, the Dutch glasshouse sector is often the best source of data on the production and trade in protected crops in Europe and worldwide. Though the Netherlands itself is a small country with a population of about 15 million, it is the acknowledged world leader in protected crop technology, production and international horticultural trade.

[3] Thus, another great advantage of the European produce auctions for growers and buyers is their objective and professional quality inspection before every transaction. Such a system eliminates buyers' dependence on growers to determine product quality when setting prices and ends growers' dependence on buyers to assess product quality upon receiving the produce.

[4] In any case, there is a considerable risk involved with aiming production and exports exclusively at peak holiday periods, since production and distribution cannot always be precisely timed and since other suppliers can flood the market during a given holiday, leading to substantially lower prices in a particular year than what might normally be expected.

[5] The best example of a successfully sustained protected horticultural industry is the Netherlands. Dutch glasshouse production is characterized by the predominance of relatively small, specialized, owner-operated production units (i.e. family farms), usually of one hectare or less. The grower is then free to concentrate almost exclusively on producing high-quality vegetables and flowers. However such production specialization can only exist if the overall industry is large and developed enough for other firms or public organizations to specialize in providing the other necessary support functions, such as research and extension advice, new product development, young plant propagation, production input supply, inspection services, market information and sales and distribution (Ministry of Agriculture-1992). But even in countries where it is not possible to duplicate Dutch production conditions, a smaller-scale operation will allow a larger investment per square meter in quality-enhancing technology and a greater specialization in high-quality production.

[6] Supplies from January to June 1993 remained about the same as in 1992, while prices have increased (Vakblad), showing some signs of market recovery. Nevertheless, the European flower trade is now much more cautious about the future than it was prior to 1992.

[7] One study concludes that, since Spain and the Canary Islands were among the countries most frequently found in violation of reference prices for tomatoes and other products, these policies have served as a significant barrier to the expansion of low-cost Spanish produce flows into the EC. Therefore, the authors maintain that Spain has not yet benefitted significantly from EC membership and still has enormous potential to expand trade in the future once the restrictive offer price mechanism is removed. Such an expansion of Spanish exports could displace third country (i.e. non-EC) produce exports. The rapid growth in Spanish and Canary Island tomato exports to the U.K. in the spring and summer of 1992 at low prices is cited as evidence that the accelerated phase-in of Iberian membership was already beginning to affect the market (Ritson and Swinbank).

REFERENCES – ANNEX

Adamczak, Jean, "Japan: Land of the Rising Flower Producers", *Florist,* April 1991, pp. 61-63.

Agrexco, Tel-Aviv, personal communication.

AIPH, *Annual Statistical Yearbook,* University of Hannover, Germany, various issues.

Alvensleben, Reimar von and Thomas Meier, "The Influence of Origin and Variety on Consumer Perception", Workshop on Measuring Consumer Perception of Internal Product Quality, *Acta Horticulturae,* Number 259, July 1990.

American Farm Bureau, "Volume IV: Fruit and Vegetable Issues", *NAFTA (North American Free Trade Agreement): Effects on Agriculture.* AFB Research Foundation, 1992.

Barendse, H., Unpublished Manuscript, Central Bureau of Dutch Vegetable Auctions (CBT), 1987.

Batt, Peter J., *A Strategic Evaluation of the Export Flower Industry in Australia,* Curtain University of Technology, Perth, January 1993.

Blauw, J., "New Developments in Flowers", VBA, lecture in Tel-Aviv, April 22, 1993.

Brooker, J., D.B. Eastwood and R.H. Orr, "Consumers' Perception of Locally Grown Produce at Retail Outlets", *Journal of Food Distribution Research,* February 1987, pp. 99-107.

Brumfield, Robin G. and Adesoji O. Adelaja, "An Analysis of Consumers' Purchasing Patterns, Perceptions, and Expenditures on Fresh Tomatoes in New Jersey", *Acta Horticulturae,* Number 295, May 1991.

Burg, S.P., "Hypobaric Storage and Transportation of Fresh Fruits and Vegetables", *Symposium: Postharvest Biology and Handling of Fruits and Vegetables,* Avi Publishing Co., Westport, Connecticut, 1975.

Cathey, H. Marc, "Creating a Market-Driven Research Plan for the `Green Industries'", International Workshop on New Frontiers for a Market-Driven Horticultural Industry, Tel-Aviv, June 1993.

De Kreij, Ronald, and Geert van Oosterhout, "Europese Tomatenmalaise", *Groenten + Fruit/Algemeen,* No. 36, 10 September 1993, pp. 14-18.

Deters, St., "Analyse der Verbraucher—und Handlerprferenzen bei Frischgemse", *Forschungsberichts zur konomie im Gartenbau,* Heft 55, Hannover und Weihenstephan, 1985.

Doesburg, J. van, "Lecture for the Flower Board of Israel", VBN, lecture in Tel-Aviv, March 25, 1992.

Egberts, Eric, "Japanese Floriculture Combines Tradition and Change", *FloraCulture International,* November/December 1992, pp. 14-17.

Elhanan, Shmuel, "The Horticultural Industry Towards the Requirements of the E.E.C. in Plant Protection and Production Methods", International Workshop on New Frontiers for a Market-Driven Horticultural Industry, Tel-Aviv, June 1993.

Euromonitor, *Market Direction Report 1.16 - Fruit and Vegetables,* London 1992.

Fruit and Vegetable Markets, statistical tables, various issues.

Fruit and Vegetable Markets, "EC Concern at Moroccan Tomato Expansion", *Fruit and Vegetable Markets,* April 1993, pp.11-12.

Fruit and Vegetable Markets, "Problems for Moroccan Export Growers", *Fruit and Vegetable Markets,* July 1993, p. 15.

F-S (Federal-State) Market News, "Ornamental Crops National Market Trends," USDA, San Fransisco, CA, various issues.

Groenten + Fruit/Algemeen, "Marktoverzicht tomaat", *Groenten + Fruit/Algemeen,* No. 50, 11 December 1992.

Grower, "Dutch Colour Up Tomatoes", *Grower,* June 24, 1993, pp. 6-7.

Hendrick, Pieter, "Dutch Auctions: From a Countervailing Power to a Market-Oriented Sales Organization", International Workshop on New Frontiers for a Market-Driven Horticultural Industry, Tel-Aviv, June 1993.

IFB (Israel Flower Board), Tel-Aviv, personal communication.

IFW (International Fruit World), "Green Light for Exports to Japan!" *International Fruit World,* 1-1993, pp. 231-232.

IKC, *Kwantitatieve Informatie voor de Glastuinbouw,* Netherlands Ministry of Agriculture, Nature Management and Fisheries, 9th edition, September 1991.

ITC, *Floricultural Products: A Study of Major Markets,* International Trade Centre, UNCTAD/GATT, Geneva, 1987.

Jaffee, Steven, "Marketing Africa's Horticultural Exports: A Transaction Cost Perspective", Paper presented at the Workshop on the Globalization of the Fresh Fruit and Vegetable System, University of California, Santa Cruz (December 6-9, 1991)

"How Private Enterprise Organized Agricultural Markets in Kenya," World Bank PRE Working Paper, 1992)

JETRO-F, "The Japanese Market for Cut Flowers", *Tradescope,* April 1992, pp. 7-18.

JETRO-Q, "New Plant Quarantine Guide Released", *Tradescope,* May 1992, pp. 19-21.

JETRO-V, "The Japanese Market for Fresh Vegetables", *Tradescope,* February 1992, pp. 7-18.

Kijne, A.G., M. Mulder and V.C. Bouwman, *De Snijbloementeelt Onder Glas, 1975-1995,* LEI, PR.NO. 41-92, The Hague, July 1992.

LEI-CBS, *Tuinbouwcijfers 1992,* Landbouw Economisch Instituut/Centraal Bureau voor de Statistiek, s'Gravenhage, The Netherlands, Juni 1992.

Love, John M., "The Produce Industry's Challenge to Double Demand for Fresh Vegetables", *Vegetables and Specialties: Situation and Outlook Yearbook,* USDA-ERS, Washington, D.C., December 1991, pp.67-70.

Lichine, Alexis, "Appelation d'Origine Controlee (Control Laws)", *Encyclopedia of Wines,* A.A. Knopf Ltd., New York 1982, pp. 72-75.

Malter, A., "1991/92 Export Season for Cut Flowers", Ministry of Agriculture, Market Research Department, Tel-Aviv, July 1992.

Melamed, Abraham, "Linking Agricultural Production to the Market", Ministry of Agriculture, Market Research Department, Tel-Aviv, July 1993.

Miller, M., George J. Ball Inc., personal communication.

Ministry of Agriculture-1990, *Essentials: Multi-Year Crop Protection Plan*, Netherlands Ministry of Agriculture, Nature Management and Fisheries, The Hague, August 1990.

Ministry of Agriculture-1992, *Essentials: Glasshouse Vegetable Growing in the Netherlands*, Netherlands Ministry of Agriculture, Nature Management and Fisheries, The Hague, May 1992.

Neff, Jack, "Produce Under Glass: What Is Its Potential for Retail Growth?" *Produce Business*, pp.14-18, June 1992.

Nienhuis, J., P. Vermeulen and G. Stiekema, "Rentabiliteit 1992 Keldert naar Dieptepunt", *Groenten + Fruit/Glasgroenten*, No. 50, 11 December 1992, pp. 12-17.

Nunnink-A, Edwin, "Ook Zo Geschrokken?" *Groenten + Fruit/Glasgroenten*, No. 14, 9 April 1993, pp. 15-19.

Nunnink-B, Edwin, "Die verrekte Moren toch", *Groenten + Fruit/Glasgroenten*, No. 19, 14 Mei 1993, pp. 12-13.

Ohkawa, Kiyoshi, "Advances in Japanese Ornamental Horticulture", *Chronica Horticulturae*, Vol. 33, Number 1, April 1993, pp.3-4.

Pertwee, Jeremy, *International Floriculture Trade Statistics 1993*, Pathfast Publishing, U.K., March 1993.

PGF, *Jaarverslag*, The Hague, various years.

PGF, *Markt-Info*, The Hague, various issues.

Porter, Ed, "U.S. Fresh Vegetable Exports", *Horticultural Products Review*, USDA-FAS, August 1992, pp. 28-30.

Prevor, Jim, "Three Cheers for Shrink!" *Floral Business*, October 1991.

PVS, *Jaarverslag*, The Hague, various issues.

Rabobank-F, *A View of International Competitiveness in the Floristry Industry*, Rabobank Nederland, June 1992.

Rabobank-V, *International Competitiveness in the Vegetable-growing Sector*, Rabobank Nederland, September 1992.

Ritson, Christopher and Alan Swinbank, *Prospects for Exports of Fruit and Vegetables to the European Community After 1992*, FAO, Rome 1993.

Seker, Suzet, "Trip Report: High-Quality Tomatoes", Ministry of Agriculture, Market Research Department, Tel-Aviv, September 1992.

Smith, John, "The Changing Floriculture Industry: A Statistical Overview by the Society of American Florists Statistics Committee", Society of American Florists, 1993.

Somers, Joe and Mark Thompson, "Japan Is No.1 and Growing Market for U.S. Horticulture", *AgExporter*, September 1992, pp. 18-21.

Staal, P. *Market Survey: Spray Carnations*, Ministry of Agriculture, Market Research Department, Tel-Aviv, February 1988.

Sugiyama, S., Flora International Co. Ltd., Tokyo, personal communication.

Thompson-A, Mark, "Japanese Imports of Horticultural Products from the United States and the World, 1989-1991", *Horticulture Products Review*, USDA-FAS, April 1992, pp. 12-14.

Thompson-B, Mark, "North American Free Trade Agreement Impact on Horticulture", *Horticulture Products Review*, USDA-FAS, September 1992.

Thompson-C, Mark, "European Community Imports of Horticultural Products, 1992", *Horticultural Products Review*, USDA-FAS, Washington, D.C., August 1993, pp. 9-18.

Tradstat, Monthly International Trade Statistics, U.K.

USDA-AMS, *Fresh Fruit and Vegetable Arrivals in Eastern Cities*, USDA-AMS, Fruit and Vegetable Division, Washington, D.C., 1992.

USDA-AMS, *Fresh Fruit and Vegetable Arrivals in Western Cities*, USDA-AMS, Fruit and Vegetable Division, Washington, D.C., 1992.

USDA-ERS, *Vegetables and Specialties: Situation and Outlook Report*, Washington, D.C., various issues.

USDA-FAS, *Horticultural Products Review*, Washington, D.C., March 1993.

USDA-NASS, *Floriculture Crops: 1992 Summary*, Washington, D.C., April 1993.

Vakblad-A, "Marktaandeel in Japan Daalt", *Vakblad voor de Bloemisterij*, No. 18, 1993, p.12.

Vakblad-B, "Goede juni-omzet leidt tot redelijke halfjaarcijfers bloemenveilingen", *Vakblad voor de Bloemesterij*, No. 27, 1993, p.8.

Vakblad voor de Bloemisterij, various issues.

VBA, "Statistiekboek exit", *Aalsmeer Nieuws*, February 19, 1993.

VBN, *Statistiekboek 1992*, Association of Dutch Flower Auctions (VBN), Leiden, The Netherlands, January 1992.

Vesseur, W.P. "Tomato Tasting and Consumer Attitude", *Acta Horticulturae*, Number 259, July 1990.

Distributors of World Bank Publications

ARGENTINA
Carlos Hirsch, SRL
Galeria Guemes
Florida 165, 4th Floor-Ofc. 453/465
1333 Buenos Aires

Oficina del Libro Internacional
Alberti 40
1082 Buenos Aires

**AUSTRALIA, PAPUA NEW GUINEA,
FIJI, SOLOMON ISLANDS,
VANUATU, AND WESTERN SAMOA**
D.A. Information Services
648 Whitehorse Road
Mitcham 3132
Victoria

AUSTRIA
Gerold and Co.
Graben 31
A-1011 Wien

BANGLADESH
Micro Industries Development
 Assistance Society (MIDAS)
House 5, Road 16
Dhanmondi R/Area
Dhaka 1209

BELGIUM
Jean De Lannoy
Av. du Roi 202
1060 Brussels

BRAZIL
Publicacoes Tecnicas Internacionais Ltda.
Rua Peixoto Gomide, 209
01409 Sao Paulo, SP

CANADA
Le Diffuseur
151A Boul. de Mortagne
Boucherville, Québec
J4B 5E6

Renouf Publishing Co.
1294 Algoma Road
Ottawa, Ontario
K1B 3W8

CHINA
China Financial & Economic
 Publishing House
8, Da Fo Si Dong Jie
Beijing

COLOMBIA
Infoenlace Ltda.
Apartado Aereo 34270
Bogota D.E.

COTE D'IVOIRE
Centre d'Edition et de Diffusion
 Africaines (CEDA)
04 B.P. 541
Abidjan 04 Plateau

CYPRUS
Center of Applied Research
Cyprus College
6, Diogenes Street, Engomi
P.O. Box 2006
Nicosia

CZECH REPUBLIC
National Information Center
P.O. Box 668
CS-11357 Prague 1

DENMARK
SamfundsLitteratur
Rosenoerns Allé 11
DK-1970 Frederiksberg C

DOMINICAN REPUBLIC
Editora Taller, C. por A.
Restauración e Isabel la Católica 309
Apartado de Correos 2190 Z-1
Santo Domingo

EGYPT, ARAB REPUBLIC OF
Al Ahram
Al Galaa Street
Cairo

The Middle East Observer
41, Sherif Street
Cairo

FINLAND
Akateeminen Kirjakauppa
P.O. Box 128
SF-00101 Helsinki 10

FRANCE
World Bank Publications
66, avenue d'Iéna
75116 Paris

GERMANY
UNO-Verlag
Poppelsdorfer Allee 55
53115 Bonn

GHANA
Greenwich Mag. and Books
Rivera Beach Hotle
PO Box 01198
Osu-Accra

GREECE
Papasotiriou S.A.
35, Stournara Str.
106 82 Athens

HONG KONG, MACAO
Asia 2000 Ltd.
46-48 Wyndham Street
Winning Centre
7th Floor
Central Hong Kong

HUNGARY
Foundation for Market Economy
Dombovari Ut 17-19
H-1117 Budapest

INDIA
Allied Publishers Private Ltd.
751 Mount Road
Madras - 600 002

INDONESIA
Pt. Indira Limited
Jalan Borobudur 20
P.O. Box 181
Jakarta 10320

IRAN
Kowkab Publishers
P.O. Box 19575-511
Tehran

IRELAND
Government Supplies Agency
4-5 Harcourt Road
Dublin 2

ISRAEL
Yozmot Literature Ltd.
P.O. Box 56055
Tel Aviv 61560

R.O.Y. International
P.O.B. 13056
Tel Aviv 61130

ITALY
Licosa Commissionaria Sansoni SPA
Via Duca Di Calabria, 1/1
Casella Postale 552
50125 Firenze

JAMAICA
Ian Randle Publishers Ltd.
206 Old Hope Road
Kingston 6

JAPAN
Eastern Book Service
Hongo 3-Chome, Bunkyo-ku 113
Tokyo

KENYA
Africa Book Service (E.A.) Ltd.
Quaran House, Mfangano Street
P.O. Box 45245
Nairobi

KOREA, REPUBLIC OF
Pan Korea Book Corporation
P.O. Box 101, Kwangwhamun
Seoul

Korean Stock Book Centre
P.O. Box 34
Yeoeido
Seoul

MALAYSIA
University of Malaya Cooperative
 Bookshop, Limited
P.O. Box 1127, Jalan Pantai Baru
59700 Kuala Lumpur

MEXICO
INFOTEC
Apartado Postal 22-860
14060 Tlalpan, Mexico D.F.

NETHERLANDS
De Lindeboom/InOr-Publikaties
P.O. Box 202
7480 AE Haaksbergen

NEW ZEALAND
EBSCO NZ Ltd.
Private Mail Bag 99914
New Market
Auckland

NIGERIA
University Press Limited
Three Crowns Building Jericho
Private Mail Bag 5095
Ibadan

NORWAY
Narvesen Information Center
Book Department
P.O. Box 6125 Etterstad
N-0602 Oslo 6

PAKISTAN
Mirza Book Agency
65, Shahrah-e-Quaid-e-Azam
P.O. Box No. 729
Lahore 54000

PERU
Editorial Desarrollo SA
Apartado 3824
Lima 1

PHILIPPINES
International Book Center
Suite 1703, Cityland 10
Condominium Tower 1
Ayala Avenue, H.V. dela
 Costa Extension
Makati, Metro Manila

POLAND
International Publishing Service
Ul. Piekna 31/37
00-677 Warszawa

PORTUGAL
Livraria Portugal
Rua Do Carmo 70-74
1200 Lisbon

SAUDI ARABIA, QATAR
Jarir Book Store
P.O. Box 3196
Riyadh 11471

SLOVAK REPUBLIC
Slovart G.T.G Ltd.
Krupinska 4
P.O. Box 152
852 99 Bratislava 5

**SINGAPORE, TAIWAN,
MYANMAR,BRUNEI**
Gower Asia Pacific Pte Ltd.
Golden Wheel Building
41, Kallang Pudding, #04-03
Singapore 1334

SOUTH AFRICA, BOTSWANA
For single titles:
Oxford University Press
 Southern Africa
P.O. Box 1141
Cape Town 8000

For subscription orders:
International Subscription Service
P.O. Box 41095
Craighall
Johannesburg 2024

SPAIN
Mundi-Prensa Libros, S.A.
Castello 37
28001 Madrid

Librería Internacional AEDOS
Consell de Cent, 391
08009 Barcelona

SRI LANKA AND THE MALDIVES
Lake House Bookshop
P.O. Box 244
100, Sir Chittampalam A.
 Gardiner Mawatha
Colombo 2

SWEDEN
Fritzes Fackboksforetaget
Regeringsgatan 12, Box 16356
S-106 47 Stockholm

Wennergren-Williams AB
P. O. Box 1305
S-171 25 Solna

SWITZERLAND
Librairie Payot
Case postale 3212
CH 1002 Lausanne

Van Dierman Editions Techniques - ADECO
P.O. Box 465
CH 1211 Geneva 16

TANZANIA
Oxford University Press
Maktaba Street
P.O. Box 5299
Dar es Salaam

THAILAND
Central Department Store
306 Silom Road
Bangkok

TRINIDAD & TOBAGO
Systematics Studies Unit
#9 Watts Street
Curepe
Trinidad, West Indies

UGANDA
Gustro Ltd.
1st Floor, Room 4, Geogiadis Chambers
P.O. Box 9997
Plot (69) Kampala

UNITED KINGDOM
Microinfo Ltd.
P.O. Box 3
Alton, Hampshire GU34 2PG
England

ZAMBIA
University of Zambia Bookshop
Great East Road Campus
P.O. Box 32379
Lusaka

ZIMBABWE
Longman Zimbabwe (Pvt.) Ltd.
Tourle Road, Ardbennie
P.O. Box ST 125
Southerton
Harare